Jürgen Weber **The Judgement of the Eye**

Jürgen Weber

The Judgement of the Eye

The Metamorphoses of Geometry - One of the Sources of Visual Perception and Consciousness

(A Further Development of Gestalt Psychology)

SpringerWienNewYork

LIST OF CONTENTS

PREFACE

This book is essentially about the question of what forms say to us, what information they convey about their very existence, how we understand their language. How does their expression come about?

Gestalt psychology, neurophysiology and the psychology of perception have hitherto tried to answer the question of whether we see forms as a whole or as the sum of their parts, why as a rule we perceive things as they actually are and not as they appear on the retina, which is in fact changed in perspective and in size, and which visual cortical areas and which neurons react to which phenomena. But the most important question for me is what the thing that is seen informs us additionally.

How do we tell the difference between a cheerful and a gloomy face? Why do we see that a bud will open shortly? Why do we find some phenomena to be dangerous and others to be desirable?

This question has not yet been investigated in a systematic and scientific way, although it is of vital importance to our behaviour, to our attitude towards the things we see. Apart from this, it is able to bring together the results of various disciplines and answer many an old question, e.g. the one about spatial, three-dimensional perception.

This is, of course, the most central issue for an artist, who, after all, wants to inform his fellow beings about the forms he has invented. The essential issue for him is to do with his creations' expression. In this respect no-one has undertaken as many experiments as he. A good many discussions with others could also be of help though:

Firstly and most importantly I talked to my father, H. H. Weber, the former Director of the Max Planck Institute for Physiology in Heidelberg. These discussions were often controversial, as are my talks on the same subject with the current Director of this Max Planck Institute, Bert Sakmann. These came about when I was doing his portrait for the Gallery of the Max Planck Society.

I gained a great deal from these conflicting views on the one hand, but on the other hand they also helped me to make my own thoughts more precise. I am grateful to both men for this.

For more than 20 years I maintained a constant dialogue with Rudolf Arnheim. Over and above the literature on gestalt psychology these very discussions brought me close to the essence of this academic field.

Thanks are due to many others, for instance the psychologist Professor Ernst Pöppel and Professor Singer, the Director of the Max Planck Institute for Brain Research, for certain ideas and not least to my assistants and students for their questions.

I would particularly like to thank the psychologist Dr Klaus Nippert of the Technical University of Braunschweig, who patiently filled the gaps in my knowledge of psychology again and again. I am grateful to him for many literary references.

Many people have worked on this book, translators, secretarial staff. I would like to express my thanks to them for patiently revising my constantly changing texts, so that finally, after ten years, this, my second work, on this subject, has been written.

I would be pleased if my reflections about a function of visual perception, perceptual judgement, my experience as an artist would stimulate some related academic fields.

Braunschweig, April 2002

PART I

Geometric Concepts of the Visual Cortex as the Basis of Visual Information

Chapter 1

Short Summary of the Main Ideas

My theory about perception on a phenomenological basis is unlike most of the work produced in this field. It does not primarily centre on the question of why we see things as they actually are and not as they appear on the retina, and whether we see them first as a whole – simplified geometrically – or whether we construct the form from details. Nor is it based on the sequence of events to be considered when seeing, hearing and action occur. In contrast, my theory centres on the question of why the things in our environment inform us of certain facts visually, or to put it more simply, how they arrive at their particular expression.

This theory joins the theories of the gestalt psychologists, with those of the Lie Transformation Groups in their applications to visual perception, my own experience and to a certain extent also neurophysiological findings. To put this differently, it harmonizes parts of these theories with each other and resolves existing differences and problems.

Both the gestalt psychologists, who approach their subject from a phenomenological basis, and the Lie Group with their mathematical approach, build on the hypothesis that forms of Euclidean geometry are of great importance for seeing and memory – for the ability to identify an object visually as well as for the visual "declaration memory" and short-term memory. Other theories have been developed recently, based on explaining visual perception on a fractal basis.[1]

Neurophysiology has meanwhile produced an amazingly exact atlas of the brain and the workings of the neurons, but basically it can only contribute very little to these discussions.

Gestalt psychology and the Lie Transformation Group (Hoffman and Dodwell, among others) believe that in the seeing process a certain agreement can be detected between the phenomena in our environment and the so-called Euclidean geometric forms or the corresponding "orbits" of the Lie Group. The weak point of these theories is that an infinitely large diversity of phenomena have to coincide with a comparatively small group of Euclidean forms to allow conclusions to be drawn. This is unlikely and has thus led to a stagnation in both of these fields.

My theory states that we do not perceive, judge and remember something by producing a correspondence between the Euclidean forms or "orbits" and the phenomena in our environment, but rather that we come to conclusions just because of the differences to these Euclidean forms. This theory has been supported by a large number of tests carried out on very many people, tried out using many natural phenomena from our surroundings and documented with a large amount of art work from the beginnings of mankind's different cultures, as well as an interpretation of children's drawings, the drawings of people with visual disabilities and research into infants' perception. We perceive through comparison, through comparison with the geometric forms invented by our visual cortex and the schemata made up of them. Our brains have developed these geometric forms from the basic processes of contraction and expansion, horizontal and vertical distortion and rotation. In my opinion, the invention of the geometric forms from the basic processes is inherited with our genes.

The same fundamental events then take place in a second process in which these basic geometric forms are distorted again, or we recognize natural phenomena as the result of such distortions. As the Euclidean basic forms can be distorted in an endless variety of ways, we can thus assess the diversity of phenomena in our surroundings.

The identification memory – that is the recognition of things we have already seen – is found in a large part of the animal world and is similar to that memory in humans. The beginnings of "declaration memory" can also be seen in the animal kingdom, especially in primates, but humans are distinguished from all the animals by an enormous advance in this field. In my opinion consciousness is directly dependant on the "declaration memory" and by no means only or primarily on the ultra short-term memory[2]. Attentiveness, corporeal awareness, sexuality and many other things are, of course, also a part of consciousness. However, nothing shapes our memory in my

[1] Jan Drösler (lecture in Braunschweig)

[2] Francis Crick "The Astonishing Hypothesis: The Scientific Search for the Soul". London, 1994

opinion as much as the "declaration memory" the visual part of which I have named "reproduction memory". Those who are able to describe the way that the reproduction memory functions, can describe an important part of our consciousness. And those who are able to describe how the reproduction memory continues to learn, can make a significant contribution to the maturing of human consciousness.

These theories of visual perception cannot only be the basis of training our perception, of new developments in this field, but can also be of great help to artists.

Chapter 2
What Is Seeing? How Visual Memory Is Affected by Agnosia and Alzheimer's Disease

Almost half of our brain – about 27 cortical areas – is concerned with visual perception. Everyone needs to be able to see the world correctly and orient himself in it in order to survive. If the number of visual areas or the amount of visual modules[3] are anything to go by, seeing and understanding must be one of the most complex happenings in our brains, though we never notice how complicated it is. If we are to use them properly, we have to put a conscious effort into learning walking, swimming, logical thought, sensible judgement, languages, mathematics, sciences and many other things, but nothing, apparently, could be simpler than seeing. We seem to be born with the ability; everyone can see. So why bother to think about it?

In his book "An Anthropologist on Mars", published in 1995, the American neurologist Oliver Sacks[4] describes in the essay "To See and Not See" the case of a fifty-year-old man who was almost blind and could only distinguish between light and dark and determine the direction from which light was coming. He had almost totally lost his sight between the ages of six and seven and so had spent more than 40 years as a blind man, learning to find his way around by using a stick, feeling objects and depending on his hearing. He earned his living as a massage therapist.

In his fiftieth year an operation restored most of his sight. When the bandages were removed after the operation he later said that, "in this first moment he had no idea what he was seeing". He perceived light, movement, colours, a hazy, meaningless muddle, and not until a voice emerged from this chaos saying, "Well?", did he realize that this chaos of light and shadows was a face – the face of his surgeon.

This patient's experience was almost identical with that of earlier patients who had gone blind as children and only regained their sight years later, after an operation. Oliver Sacks writes, "The rest of us, born sighted, can scarcely imagine such confusion. For we, born with a full complement of senses, and correlating these, one with the other, create a sight world from the start, a world of visual objects and concepts and meanings. When we open our eyes each morning, it is upon a world we have spent a lifetime learning to see. We are not given the world: we make our world through incessant experience, categorization, memory, reconnection."

Patients who have always been blind get used to experiencing the world by touch. Just think what that means. They could not perceive space as three-dimensional, but only in one-dimensional time. They know how many steps to take between the bed and the door, how many there are from home to work etc. Besides this, and this is almost more important, they have to touch the objects around them. For instance, touch a head and it is always identical with itself. They feel a forehead, eyes, temples, cheekbones, a nose, mouth, chin and neck and it stays the same. But when Virgil – that is the name of the patient I mentioned – opened his eyes, he saw a human face divided up by light, shade and movement. (1, 2, page 118) To illustrate this idea I would like to present two Rembrandt self-portraits, painted around 1630. They show the same head, at the same age, but differently lit and in different poses. Bright patches of light, shadows and demi-shadows are different in both cases, and yet we recognize the same man in both pictures, despite different areas of light and shade and the head facing in different directions. What we have to ask ourselves is, how does this occur?

[3] Ernst Pöppel "Eine neuropsychologische Definition des Zustandes Bewußt", in "Gehirn und Bewußtsein". Weinheim, 1989
[4] Oliver Sacks "An Anthropologist on Mars: Seven Paradoxical Tales". New York, 1995

Virgil had a dog and a cat and while he was blind he could easily feel which was which. However, he could not do this by looking at them. Both were black and white, so he constantly muddled them up, because he simply could not grasp their total appearance by using his eyes. While he was blind, each animal was identical with itself: he felt each animal's body, head, four legs and tail and each animal's coat and skeleton felt different. When he looked at them, however, he was not able to identify one animal from the other through their movement, light and shade. (3) Depictions of the different movements of the same cat show how very different the cat can look. We recognize the same animal as it moves about, but someone who has been blind for years has never learnt to do so.

In the optical world, objects are superimposed upon each other. A blind man, feeling his way through the world, never experiences this. (4) The Descent from the Cross, a late Rembrandt etching dated 1655, is a wonderful example of all three phenomena. The objects are changed by light and shade, dissolve into patches of varying degrees of brightness, movement misshapes the human form in widely differing ways. The grouping superimposes figures and objects upon each other. Despite this, we realize that the lower legs and head of the horizontal figure belong to Christ, lowered from the cross, carried and overlapped by Nicodemus, seen from the back. We can certainly recognize the bright form stretching up vertically to the top of the picture as a combination of illuminated shroud and assistants. We can connect the raised arm emerging from the dark with the scarcely visible human form. Unlike the blind, we have "learnt" to analyze, understand and judge our constantly changing visual world.

Nor could Virgil use his eyes to understand simple, unmoving geometrical forms, which had not been altered by lighting effects, and which he had recognized easily by touch when still blind. Circle, cube (5), rectangle and triangle – altered in perspective – made no sense to the seeing Virgil, even though while still blind he had learnt the feel of these basic geometric forms. But now, in visual reality, circle, cube, rectangle and triangle were not identical with themselves. From side views a circle looked like various kinds of ellipse, a cube like different unsymmetrical geometric forms, rectangles like all sorts of trapezium, right angles became obtuse or acute. To help Virgil at least to recognize these basic forms, he was given a box of wooden spheres, rectangles and triangles which had to be fitted into the correct holes. It took him four weeks to identify these basic geometric forms visually, without touching them as well. But even later he preferred touching them.

He continued to point out that he no longer felt confident, and that he could orient himself better in his blind world, using touch and his stick, than he could now with his limited visual function. All the time he worried about making mistakes and wanted to return to the world of the blind. He did not only lack a sighted person's visual experience, but – directly connected to this – many anatomical preconditions in his brain, neuron connections etc., in order to collect enough new visual impressions after his operation.

Shortly afterwards he contracted a severe case of pneumonia and suffered from an acute lack of oxygen. This caused him to lose his sight again and he returned contentedly, even with relief, to the world of the blind.

This medical history gives us an idea of the complicated process involved in decoding our visual impressions into meaning, expression, spatial order and function. We learn of similar post-operative results after the successful treatment of other patients who have been blind for many years. Similar results have also been observed when brain damage causes people to lose the ability to see forms.

Even though in recent decades great progress has been made in anatomical, chemical and electrophysiological research into the cortical areas of the brain and neurons, it is still only possible to answer these questions inexactly. Researchers now know what stimulates which brain cells, so that they cannot only differentiate between the visual and aural cortices, but in many cases they can also determine, for instance, which parts of the visual cortex react chemically and electrically to which visual stimuli. But so far it has not been possible to discover why stimulating one cell causes acoustic impressions, while stimulating another causes visual ones, nor yet how these visual impressions may be responsible for connected, significant pictures. So ideally researchers should be working on the question simultaneously from two sides – and both are equally important – they should be working towards each other, as if drilling a tunnel from opposite ends, from the chemical, physiological/electrical side and from the phenomenological side. It is likely, that the results of these two very different areas of research could prove mutually beneficial, or at the very least, lead to questions which would not occur independently to either method. I do not think, like Crick[5], that the starting point should almost exclusively be neurons and their interactions when answering these questions.

Besides this, the visual perception model suggested here could solve one of the central questions of prehistoric research. It could provide prehistoric and early historic research with important indicators and prevent the development of false theories. But before we examine this in detail, here is another of Oliver Sacks's stories.

In his book "The Man Who Mistook His Wife for a Hat"[6], published in 1985, Oliver Sacks writes, "Dr P. was a musician of distinction, well-known for many years as a singer, and then, at the local School of Music, as a teacher. It was here, in relation to his students, that certain strange problems were first observed. Sometimes a student would present himself, and Dr P. would not recognize him; or, specifically, would not recognize his face. The moment the student spoke, he would be recognized by his voice. Such incidents multiplied, causing embarrassment, perplexity, fear – and, sometimes, comedy. For not only did Dr P. increasingly fail to see faces, but he saw faces when there were no faces to see: genially, Magoo-like, when in the street, he might pat the heads of water-hydrants and parking-meters, taking these to be the heads of children; he would amiably address carved knobs on the furniture, and be astounded when they did not reply…"

His condition deteriorated continuously, but it did not occur to Dr P. that anything was wrong until about three years later, when he developed diabetes. Since he knew that this illness can affect the eyes, he consulted a specialist, who examined him thoroughly. Finally the doctor said, "There's nothing the matter with your eyes,… But there is trouble with the visual parts of your brain. You don't need my help, you must see a neurologist."

Dr P. increasingly lost the ability to recognize objects as a whole. When presented with pictures, "His eyes would dart from one thing to another, picking up tiny features, individual features,... A striking brightness, a colour, a shape would arrest his attention and elicit comment – but in no case did he get the scene-as-a-whole. He failed to see the whole, seeing only details, which he spotted like blips on a radar screen." (This is how visual perception was viewed in the nineteenth-century. The author)

Yet Dr P. still combined "a perfect ear and voice with the most incisive musical intelligence… Dr P.'s temporal lobes were obviously intact: he had a wonderful musical cortex." But "what … was going on in his parietal and occipital lobes, especially in those areas where visual processing occurred?" Without sound he was unable to identify his optical, natural surroundings. "He failed to identify the expressions" on faces. When shown a film with the sound off, for instance, "He was very unclear as to what was going on, or who was who or even what sex they were. His comments on the scene were positively Martian." as Oliver Sacks tells us.

In short, he no longer recognized anyone, not even his family. He recognized Einstein by his typical moustache and hair. He approached pictures of people – even his nearest and dearest – as if they were abstract puzzles or tests. For him, each face was a new collection of elements. In her book "Mapping the Mind", Rita Carter[7] describes a man suffering from agnosia who could only recognize his wife in company because she wore a big red bow in her hair.

As we can see, there are similarities between this and the previous story. The blind man, suddenly gifted with sight, could not put his world together visually, but only recognize individual patches. The patient whose visual centre obviously continually deteriorated within his brain, must have found himself more and more in the same position.

He was only able to identify objects by touch, or by trial and error. This did not apply to the basic geometric forms. He could still see a cube, a rec-

[5] Francis Crick "The Astonishing Hypothesis: The Scientific Search for the Soul", op. cit.
[6] Oliver Sacks "The Man Who Mistook His Wife for a Hat". 1985
[7] Rita Carter "Mapping the Mind". London, 1998

tangle, a square or a circle. Finally these were the only things he could still understand. And simple geometric forms were among the few things the blind man who recovered his sight somewhat painfully managed to learn to recognize.

However, Dr P. not only sang, he also painted. His early works were all lifelike, "naturalistic and realistic", Oliver Sacks writes, "with vivid mood and atmosphere, but finely detailed and concrete. Then, years later, they became less vivid, less concrete, less realistic and naturalistic; but far more abstract, even geometrical and cubist." Oliver Sacks goes on to say, "Finally, in the last paintings, the canvasses became nonsense, or nonsense to me – mere chaotic lines and blotches of paint. I commented on this to Mrs P. 'Ach, you doctors, you're such philistines!' she exclaimed, 'Can you not see artistic development – how he renounced the realism of his earlier years, and advanced into abstract, non-representational art?"

"No, that's not it," writes Oliver Sacks, "He had indeed moved from realism to non-representation to the abstract, but this was not the artist, but the pathology, advancing – advancing towards a profound visual agnosia, in which all powers of representation and imagery, all sense of the concrete, all sense of reality, were being destroyed. This wall of paintings was a tragic pathological exhibit, which belonged to neurology, not art."

Dr Sacks then concedes that possibly artistic and neurological development coincided. The important point for us is the realization that because of the destruction of the visual centres, the pictures gradually rejected reality in favour of basic geometric forms. Dr P. called this his "cubist period". Eventually even these geometric forms disappeared. Dr Sacks writes, "in the final pictures, I feared, there was only chaos and agnosia." We could also say, only movement. There were traces of movement not far removed from Abstract Expressionism, Art informel (16, 20, 21) or the scribbles of a two to three-year-old child.

Similar cases of agnosia have often been described. What makes this case almost unique is, however, the fact that the patient was also a painter. It is possible that he underwent development in reverse to that of an infant. That is what makes this particular case especially informative.

A small travelling exhibition was held in Germany some time ago[8]. It showed the development of the paintings and drawings by the famous graphic designer Carolus Horn who suffered from Alzheimer's disease and also painted in his free time. All the visually perceptive centres of patients with agnosia slowly deteriorate. In Alzheimer's disease more and more of the centres are destroyed. Mrs Horn collected all her husband's works up until he died.

I would like to illustrate Oliver Sack's descriptions using the development of these pictures as no illustrations of the chosen case are included in his book. Agnosia and Alzheimer's both have the same kinds of effects on the cortical areas of visual perception.

At the beginning – before the disease started – Carolus Horn painted and drew, just as described by Oliver Sacks, realistic, professional landscapes (6, 7), complicated in form and colour. Then – still prior to his medical diagnosis and the corresponding symptoms – his style became increasingly coarse and geometrical (8, 9), human figures and gondolas became schematic, directions more parallel. While in figure 7 the surface of the bridge appears to be constructed and the sides of the gondola are superimposed appearing elegant and vivid, the forms in figures 8 and 9 grow increasingly geometrical and schematic. In particular, I would like to draw your attention to the gondolas and people, but also to the parallel structures of the bridge. In figure 9 the crosspieces of the windows which are actually placed horizontally within the arches are depicted diagonally to the bridge itself and the arch of the bridge is structured with parallel lines and triangles at the same distance apart.

In figure 10 – a few years later – the process of parallelism, that is alignment with rectangles, has already made much progress: the tiled floor is depicted as a chessboard almost unchanged by the use of perspective, the seated figures are twice bent at near right angles – the right angle is the simplest one – and both of the oeil-de-boeuf windows consist of a collection of circles, with radials placed at equal distances. We shall come back to this at a later stage. In contrast to the distortion of space and perspective, simple geometric forms are emphasized more in the foreground, the environ-

[8] Konrad Maurer, Ulrike Maurer, Tilde Horn and Lutz Fröhlich "Alzheimer und Kunst". Exhibition Opening Carolus Horn

ment is more and more composed of distorted geometrical shapes. A similar development can be seen in figure 11. I would like to point out the regular pattern consisting of parallel triangles behind the two schematic figures on the right or the decorated frames equally spaced out and composed of ellipses and radial crosses. At this point in time, C. Horn was having difficulty in recognizing people.

Figures 12 and 13 show that the parallelism, rows placed at equal distance and the pictures made of rectangles have progressed further. The geometrical arsenal of forms is growing smaller and smaller. The animals in figure 13, for instance, only consist of rectangles and ellipses. The four ellipse shaped clouds in figure 12 placed in a row parallel to the edge of the picture are equally spaced out and are each made of double ellipses recalling fried eggs to mind.

Figures 14 and 15 finally reflect the complete deterioration of the patient's reproductive memory. In picture 14 C. Horn tries to portray houses, in the upper line still using a combination of triangles and rectangles. These however suffer from his decreasing ability to concentrate in the following parallel lines and become more and more like rectangles or circle and ellipse-like forms. The composition is reduced to the screen.

In figure 15 he is only able to produce parallel diagonal scribbles – pure traces of movement. He thus ended up with drawings similar to those that he probably created when he was between 12 and 16 months of age (16).

Chapter 3
What Do Infants Recognize and What Do Their Visual Memories Look Like?

If parts of the process of visual perception have to be learnt and developed anatomically, this apparently happens in the earliest years of life. Since it is difficult to find out what infants really see and understand in the important first year of life, I should like first – before we go into it – to take a look at children's drawings. We can draw conclusions of existential importance from these because their development looks very similar all over the world in all children given the opportunity, from the very start or from the middle of their second year, to scribble, draw and paint.

At the beginning of their second year children begin to "scribble", as we say, using a hand, moving the whole arm, vertically dotting the paper, swinging up and down, moving vertically, diagonally and horizontally[9]. These swinging movements, which echo babies' mass movements with their arms, hands and bodies in the second half of their first year, and which the child frequently accompanies with stories which, however, have little to do with the resulting image (see 16, 17), result in zigzags and lines superimposed upon each other. The points hammered onto the paper become dotted images turning into curves. The curved lines (21) become loops (18), multiple loops, spirals, multiple superimposed circles (19), which eventually take over the entire picture. This initial phase of relatively uncontrolled scribbling, taking little account of the paper size, can take up to a year. Tirelessly and repetitively the child carries out the same movements, gradually gaining more coordination and control. This can also be seen in its ability to respect the limits of the paper and to fill it with a concentrated and clear production of scribbled figures with whirled knots, cross-hatching or zigzags, all happily combined. (20 Pollock, 18, 21 children's drawings)

At the end of the third year, sometimes not till the beginning of the fourth, these spiral or loop-like lines gradually take on more closed shapes (22), which finally end in a free and rather irregularly drawn circle-like figure. More or less simultaneously the swinging movements turn into lines crossing each other at right angles, a kind of regular pattern, or ladder (22), or (23, 24, 31 a) even symmetrical horizontal/vertical crosses. Whereas initially the lines, as traces of movement, meet at accidental angles, at the end of the third and beginning of the fourth year they tend more and more to cross at right angles (23). A rectangle (28, 31) or square can arise just as easily from a ladder-like crossing of lines (22, 29) as from an angular deformation of the circle (25, 33). So at the end of its third year the child has command of the basic forms or perceptual images circle, rectangle, lines

[9] J. Matthews "Children Drawing: Are Young Children Really Scribbling?" "Early Child Development and Care", 18, 1 – 39, 1984

which cross each other to make a ladder or regular pattern, or a simple rectangular cross. At the same time, the infant continues to scribble figures as direct traces of movement (24, 31 a).

This development which can be observed in all infants in a similar form is indeed found in all cultures. The first phase of swinging traces of movement (16) gives rise in a second phase, via the development of spool-like and regular screen-like images, to the basic geometric forms: circle, rectangle, straight line and regular pattern or screen. The child does not master these basic geometric forms by an abstracting observation of its surroundings. They develop from the movement traces of the scribbling stage in many individual phases, which we can follow as if they were in slow motion. We can trace the simple closed circle from the curved line, via the multiple rotations, the knot, the spools or spiral-like open forms. The rectangle finally emerges from irregular crossings via multiple rectangular crossings, ladders and regular patterns. The basic forms develop from the basic processes of up and down, backwards and forwards, rotation and expansion. This is not contradicted by the fact that children tell stories while they are drawing. The drawing does not illustrate the story, but accompanies it with action or gesticulating scribbling.

So if researchers have recently asked themselves whether these unidentifiable scribbles really contain a meaning adults cannot understand, this problem does not touch on what we are considering. Statistically there is no doubt that scribbling begins with zigzags, goes on to parallels, up and down and backwards and forwards, and finally, via loops and spools, develops into circles, regular patterns and rectangles. In other words basic geometric forms, or forms resembling them, develop from basic processes – we will return to the question of what these are later. In any case, it seems to be simple to set aside the disagreement as to the meaning of the scribbles. In my experience, if you wait a little while and then show infants their scribbles again, they remember nothing of the stories they told while drawing and do not recognize them.

Some of the anthropoid apes who draw reach roughly this stage. They produce rectangular crossings and some of them even draw circle-like images. But while the child goes on to increasingly complex forms and finally to a more or less schematic depiction of its environment from memory, that is to a stage where there is an obvious connection between a drawing and what it means, primates' drawings never develop any further. They do not seem to develop a real reproduction memory (visual declaration memory).

In its fourth year the child begins to combine the basic forms it has so far mastered. It joins together circles, rectangles or squares (33), draws lines across them or puts lines or dots in the circle (25, 26). It starts to structure the circle like a wheel (23) or to set radials (32, 32 a, 34) on the circumference of the circle or rectangle, creating an image like a sun. These are all combinations of basic forms and simple patterns, with no intention of illustration. For instance, these figures are certainly not meant to depict the sun, though people may implant this idea in the infant's head. They are simply rotation schemata. Even if Mrs Kellog[10] took systematism too far with her 20 basic schemata from which the human figure is finally to develop, her depiction of the basic principles was nearer to the truth than many of those who subsequently corrected her, especially if the development of children's drawing is judged from the way sight functions in adults.

The individual stages of children's drawing described here are probably an expression of the development of the mind through the self organization of cognitive structures (Singer[11]), which still proceeds even if children are not given the opportunity to draw or paint. It is, however, certain that these activities encourage this development. We shall now continue with the description of this process.

If the radials are placed around the outer circumference of the circle, the space inside it is empty and can be filled with dots, lines and circles (25, 26) until – based on the simple form of symmetry – these circles, lines or dots combine in such a way, that they resemble the face schema (27). The infant realizes that its creation can be interpreted as a human face and, after a few false starts, uses this schema more and more deliberately. The "sunbeams" at the top of the circle or rectangle

[10] R. Kellog "Analyzing Children's Art". Paolo Alto, CA, 1969
[11] Wolf Singer "Zur Selbstorganisation kognitiver Strukturen", in "Gehirn und Bewußtsein". Weinheim, 1989

turn into hair and the lower ones are lengthened into arms and legs and/or ears. Initially very wobbly and at undefined angles, they very soon organize themselves in accordance with the principle of the simple, symmetrical, rectangular form, with horizontal arms and more or less vertical legs. The first figure schema – point symmetrically constructed – has developed from the basic forms – not by abstraction from observed surroundings, but by combining the basic forms mastered in the way described above; screens, circles, radials, rectangles and lines on the basis of a rotation schema. This first human schema is generally known as a "tadpole" (32 a, 35). In the fourth and fifth year it then develops, by the child's processing perceptual experiences and including horizontals, more and more to an axially symmetrical human schema, with a head, a neck, a body, legs and arms (36, 41, 42, 43, 45).

This second human schema which succeeds the pure tadpole – if we may so classify a continuous development in infant drawing – presupposes a somewhat different attitude. While the tadpole came about almost accidentally as a result of playing with the basic forms (Arnheim "Kunst und Sehen" ("Art and Visual Perception", 1954)), the child now starts to process its first and most primitive observations. The child sees that people have a head, a neck and a body and uses the basic forms it has found to depict them. The basic forms developed from the child's brain structure are now, based on the most primitive of observations, used to construct human schemata – the centrally symmetrical image becomes an axially symmetrical figure.

At some point in the fourth or fifth year the child adds the triangle to its basic forms. Initially this is just used geometrically (40), but very soon it is employed to depict the visible environment (39), houses with triangular roofs (41), boats with triangular sails (47). So in its seventh year the child can already use the basic geometrical forms it has developed to depict a landscape, trees, houses and people (43, 44). Characteristic of these landscapes, on the whole, is that all the objects depicted are much of a size: house, roof, person, tree trunk and treetop. Obviously they are based on a screen division, a "ladder system" (43, 41). This has very little to do with the explanation so frequently mentioned in this connection, the "meaning perspective", but far more with a tendency to draw similar sizes and spaces.

This short description of the way children's drawing develops is enough for now. It can be observed all over the world, almost as if governed by a law, so we can confidently assume that it is based on a physiological process of the development of the brain. My account is based on Arnheim[12], Matthews and Kellog. I have, however, further developed these similar theories in the context of my theory of perception and my own research into children's drawing.

As the reader has certainly noticed, the development of children's drawing described above is more or less the reverse of the singer's and graphic designer's diminishing abilities to perceive. Originally their perception was normal, they could recognize everything. Then their visual perception became so simple and lost the ability to recognize individuals so much that they could no longer differentiate between faces, not even those of their nearest relatives. Eventually they could perceive only basic geometrical forms, so that they confused hydrants and parking metres with children's faces. Simultaneously their painting changed from the realistic depiction of landscapes to the composition of geometric forms and finally ended up as chaotic scribbling, like an infant's first attempts at drawing.

Of course, the two processes are not quite as analogous as they here appear. Obviously the child can recognize individual faces and situations well before it even starts to scribble. But the development of its ability to draw, that is, its ability to reproduce something from its imagination, bears comparison with the process in the opposite direction of the singer's and designer's powers after contracting agnosia or Alzheimer's respectively. It is interesting to note that their ability to see diminished synchronously with the reverse development of their painting.

It is, however, possible that the development of children's drawing starting in their second year is also related to the initial stages of a baby's perception.

This idea seems to be supported by the results of the abundant research into the beginnings of

[12] Rudolf Arnheim "Kunst und Sehen". Berlin, 1964 ("Art and Visual Perception", Berkeley, Los Angeles, London, 1954)

recognition in babies and infants in their first year. At this point we do not want to go in detail into the innumerable publications on this subject – especially as there are several good surveys of this area of research[13] – nor do we want here to reflect critically on the methodology of this research, or to go into the anatomical, neurological and physiological preconditions. We just want to consider the results of this research, since they have apparently received general recognition and no longer – or almost never – cause controversy.

The first thing a newborn baby recognizes is movement. Movement is also the first thing scribbled by a child in its second and third year. Up and down, here and there, backwards and forwards, curved and straight, this is typical of the scribbles produced into the third year.

After just a few weeks newborn babies can distinguish between curved and straight lines, and also between horizontal and vertical. Babies less than six weeks old find a chessboard pattern, that is a regular pattern, very interesting. (49) Since at this age babies need a much higher contrast threshold than adults before they can recognize something, they cannot take in a linear screen, but they can recognize the stronger contrasts of a chessboard pattern. It probably corresponds to a simple function of visual cortical areas in early childhood. And, interestingly, the first system to emerge from the up-and-down scribbles of the three-year-old child is the screen or ladder (22, 29).

At about one month babies recognize the simple contours of a form, but not its individual elements, for instance, the rather circular contour of a human head, but not the individual details of a face (51 A). At the end of its third year the infant draws a rather circular image, filling it initially – at first in a rather confused manner – with little circles or other elements (25, 26), until finally a symmetrically organized face emerges and round about the fourth year the tadpole is produced. Not until their second month do babies prefer the natural arrangement of the facial features to a confused distribution of eyes, mouth and nose, which did not bother them before (51 B, C, D). Then, in their third month, they can differentiate between their mother and a stranger. They can even recognize a colour photograph of her[14]. In some cases it takes longer to reach this stage of development.

Even though this takes a newborn baby a short time, maybe a few months, while children's reproductions need at least two years to develop, the similarities are clear. Four-month-old babies take in both the outer edges or contours of a head as well as the interior facial features. They concentrate particularly on the eyes, which are also the most striking feature of the tadpole drawn by the infant. These eyes also have the geometrical form of pure circles (35).

In the sixties, seventies and early eighties the realization of a group of neuropsychologists and mathematicians (W.C. Hoffman, Dodwell and others), based on the result of Lie mathematics, was disseminated, showing that, as a result of evolution and ecological conditions, the visual system of perception had at its command "orbits" which, primitive and easily decoded, reproduce – by means of radial patterns and concentric circles, parallels and regular patterns – expansion, contraction, rotation and parallel distortion in the frontal level (50). They believed they had found out mathematically important visual concepts of our visual areas[15]. Apparently young organisms, such as babies, but also newborn kittens, can recognize these "orbits" more easily[16].

But comparable figures are drawn by the infant at the end of its third and beginning of its fourth year, with all the spools and circles around and in which it concentrically places radials, and out of which it finally develops the tadpole (19, 23, 35). The drawings of the patient with Alzheimer's disease also become more like these "orbits", for example, the increasing parallelism (9, 10, 12, 13) or the circles with radials spaced at equal distance (10, 12), the form of the clouds with ellipses placed centrally one within the other (12), of which the inner ellipses are completely meaningless etc.

[13] Peter C. Dodwell etc. "Handbook of Infant Perception from Perception to Cognition". Florida, 1987

[14] D. Maurer and M. Barrera "Infants' Perception of Natural and Distorted Arrangements of a Schematic Face", "Child Development", 52, 196 – 202, 1981

[15] Peter C. Dodwell "The Lie Transformation Group Model of Visual Perception". Ontario, 1983

[16] Peter C. Dodwell, France E. Wilkinson, Michael W. von Grünau "Pattern Recognition in Kittens: Performance on Lie Patterns". Ontario, 1983

What is certain is that after the first six months of their lives, babies recognize configurations and the elements out of which they are constructed, if they are dealing with simple, regular, geometrical forms and symmetrical organizations. However, if in this connection, "good form" is mentioned, a concept introduced by Wertheimer in the twenties, there is every reason to take seriously the doubts often expressed concerning it. The question is, whether the "good form" is the one containing the least information. A radially symmetrical form whose position in space cannot be determined because it looks the same whichever way it is turned, contains, because of this, less information than an axially symmetrical one. Indeed, infants develop their first drawing of a person from a centrally symmetrical figure, very close to these "orbits", and then go on in their fourth year to develop from it an axially symmetrical figure, that is, a form more exactly organized in space.

The fact that babies of some seven months are able to recognize the square developed by gestalt psychologists in the twenties and thirties consisting only of four circles with rectangles cut out of them (52), shows to what an extent evolution, almost from the beginning, has anchored these simple geometrical forms in our perceptual system. A short time later, they are also able to recognize triangles (53). (Kanizsa)

Perception also implies recognition. In their first weeks or months of life babies apparently recognize geometrical forms or forms resembling them first and foremost. Of course we must realize that the development of babies' perceptual ability goes much faster and proceeds much further than the equivalent process in the case of the infant drawing and reproducing images. A one-month-old baby probably cannot distinguish purely visually between its mother and a stranger, but needs to hear her voice as well. This is presumably because it recognizes only the contour of a head, but not the facial details. But after a few months it can even recognize its mother in a colour photograph. Neither a child reproducing images from memory, nor an adult, ever reaches a comparable exactness. We shall return to this point later.

As early as three months infants can distinguish between different facial expressions, at least in the case of their mother; between smiling and frowning, between surprise and annoyance.

We must find it significant that the beginnings of perception – what a newborn baby recognizes and what is initially drawn in the second year of life – are marked by a preference for movement and then for such simple geometrical forms as straight lines, rectangles, regular patterns, spools and circles and then for centrally symmetrical and axially symmetrical images. It is remarkable that they recognize first, and later produce first, geometrical forms – just as such developments can be observed in reverse during brain damage – even though during evolution the environment of homo sapiens was anything but geometrical. Nature does not confront us with the basic geometrical forms: the rectangular screen, the circle, the square, the rectangle, the triangle etc. or only in exceptional cases. It is true to say that the world around us consists not of geometrical forms, but of highly complicated natural forms. The basic geometrical forms exist so rarely that we must regard them rather as a product of our imagination than as products of or abstractions from our environment.

We do indeed know that the higher mammals have geometrical conceptions. For instance, kittens can recognize circles faster than more complicated figures. It is not for nothing that circular-radial figures or figures resembling chessboards are used in recognition experiments with apes (54). It is probable that many mammals have a conception of a rectangle.

In the nineties Dr Birmelin, a Swabian behavioural scientist, carried out the following experiment with a domestic pig: he set up four little troughs in the four corners of a large imaginary rectangle in a sandy area. Each trough was covered with a board. The pig came along, pushed the boards aside and ate. The next step was to remove one of the troughs and its board, but the pig sought food exactly where it had stood. Further experiments to check the result showed that the pig obviously had the idea of a rectangle in its head and after feeding at the three remaining troughs, trotted over to the fourth one in search of more food – this time in vain. So this was an experiment that can be compared with the square consisting only of circles from which a rectangular segment has been removed (52).

As I have already mentioned, chimpanzees who were allowed to draw managed to get as far as producing circles from their scribblings. As early as

1921, Köhler, in his book "Intelligenzprüfungen an Menschenaffen" which examined the intelligence of anthropoid apes[17] proved that apes can recognize triangles of any type and colour. He placed a row of boxes in front of the apes, one of which contained particularly tasty food and was marked with a triangle. The experiment was repeated with a smaller or larger triangle, the triangle was completely reversed, shown as a black triangle on a white background and a white triangle on a black background, or just drawn or painted. The apes had no trouble recognizing the triangle for what it was and so always got their food. Similar results were arrived at using rats. The ability to make this transference shows at least that these animals can recognize the geometrical concept of a triangle in whatever form it appears.

Even if the different positioning of mammals' eyes in their heads causes different sight conditions, there is much to show that the visual perception of all mammals, including human beings, stems from a basic model that is varied in individual species, just as, for instance, mammals' skeletons are individually different but comparable in principle. This was pointed out as early as the middle of the nineteenth century by Charles Darwin.

Indeed it even seems to be the case not only that the optical impressions of mammals' eyes are processed cerebrally in a similar way to those of human eyes, but that even birds see in much the same way. How else can we explain why the principle of mimicry, so often employed in nature, functions in almost the same way in all creatures. Animals who steal others' eggs are also deceived by the mimicry of the colourfully spotted egg (55).

Basically mimicry has two effects. Firstly, the shell, fur or skin of the creature using mimicry to make itself invisible is made to fit in with its environment, so that its contours disappear and, secondly, the simpler form of the creature is made so indistinct by striking pigmentation that it can no longer be distinguished. It is difficult to recognize the simple triangular form of the moth (58) because our eyes are attracted to the much more complicated figure of the bright pigmentation. This second effect of mimicry is specially important for creatures which frequently change their surroundings. If it moves about a lot, a creature cannot colour its fur or skin to imitate every place it goes to, but mimicry can still make its simple form difficult to recognize. Thus the snake's ornamentation (56) is by no means identical with its background, but it is difficult for the eye to pick out the long, coiling, cylindrical body with its strong ornamentation running counter to the snake's direction, unless the snake actually moves.

The same applies to the ocelot[18] (57) and to current military camouflage. Neither of these imitates their backgrounds all the time but only very rarely. However, the patches of colour constantly make it more difficult to recognize the shape of the soldsier and the camouflage suit's patches, continually combined with new light and shade in their surroundings, visually diffuse the form. In all cases it seems that recognition of the camouflaged form is made more difficult not only by imitation of the environment, but also by the optical diffusion of the creature's simpler form. This supports the idea that most animals, plus human beings, recognize objects in the same or similar ways. Initially they grasp the whole figure as a simple shape or simplified geometrical schema. If mimicry makes this difficult, recognition is also difficult.

This poses the question: to what extent do the simple geometrical forms – the first things a newborn baby recognizes and an infant draws after the scribbling phase; the last things understood by a patient whose visual centre has been destroyed – possess basic importance for the perceptual faculties of possibly all sighted creatures?

Chapter 4
The Conclusions of Gestalt Psychology and Its Limitations

The gestalt (form) psychologists were the first to recognize this question as a central problem. Up to the end of the nineteenth century it was thought that a whole was constructed by adding up its perceived details, i.e. that perception proceeded from the individual to the general, the gestalt psychologists – hence the name they chose for themselves – proceeded from the opposite hypothesis: initially a geometrically simplified whole form is perceived and only then, in further

[17] Köhler "Intelligenzprüfungen an Menschenaffen". Berlin, 1921

[18] Robert M. McClung "Die Tarnung der Tiere". Vienna, 1978 (How Animals Hide, [Washington, 1973])

acts of seeing, the details, also geometrically simplified, of course. Thus Arnheim writes in his well-known book "Art and Visual Perception" (1954), "The experimental findings demand a complete turnabout in the theory of perception. It seemed no longer possible to think of vision as proceeding from the particulars to the general. On the contrary, it became evident that over-all structural features are the primary data of perception, so that triangularity is not a late product of intellectual abstraction but a direct and more elementary experience than the recording of individual detail. The young child sees 'doggishness' before he is able to distinguish one dog from another."

Arnheim then goes on, like Köhler, to conclude that initial recognition of this simple geometrical whole form of an object corresponds to geometrical constellations in the brain. He believes such constellations to be constructed in the brain: "If the foregoing presentation is correct, we are compelled to say that perceiving consists in the formation of 'perceptual concepts.' To the usual way of thinking this is uncomfortable terminology, because the senses are supposed to be limited to the concrete whereas concepts deal with the abstract. The process of vision as it was described above, however, seems to meet the conditions of concept formation. Vision deals with the raw material of experience by creating a corresponding pattern of general forms, which are applicable not only to the individual case at hand but to an infinite number of other cases as well. By no means should the use of the word 'concept' suggest that perceiving is an intellectual operation. The processes that have been described must be thought of as occurring within the visual apparatus."

So Arnheim thinks that the recognition of a figure comes about when geometrical constellations of the visual cortex coincide with the "raw material of experience", permitting the object to be actually perceived. Recent research concerning babies seems to confirm this belief.

Gestalt psychologists conclude that we perceive visual stimulations in the simplest possible form. The explanation for this sort of phenomena is to be found in what the gestalt psychologists call the basic law of perception. This law says that each configuration of stimuli strives to be seen in such a way that the resulting figure is the simplest possible under the given circumstances. Gestalt psychologists since Koffka (1935)[19] have also recognized in this "basic law of perception" an important reason for spatial seeing (60). I again quote Arnheim:

"Figure 60 a consists of three parallelograms. If each of them assumes a tilted position, which transforms its shape into a square, we shall see the total pattern as a cube in three dimensions rather than as the much less simple flat and irregular hexagon in the frontal plane... But not all such projections will make us see a cube. In figure 60 b the effect is considerably weaker because the symmetry of the frontal figure gives much stability to the two-dimensional version. And most observers will find it difficult to see 60 c as a transparent version of 60 a. These examples illustrate the rule formulated by Koffka in his pioneer investigation on the subject. 'When simple symmetry is achievable in two dimensions, we shall see a plane figure; if it requires three dimensions, then we shall see a solid.' Slightly rephrased, the principle asserts that whether a pattern is seen as two-dimensional or three-dimensional depends on which of the two versions produces the simpler pattern." Thus far Arnheim.

The gestalt psychologists used this example to explain very clearly by means of the "law of the simple form" – misleadingly also called "good form" – the spatial effect of perspectival distortion, although at first glance it probably seems to be clearer than it really is. We shall return to this subject.

As to the importance of the simple form for perception, Arnheim goes even further: "But if we leave the world of well-defined, man-made shapes and look around in a landscape, what do we see? A mass of trees and brushwood is a rather chaotic sight. Some of the tree trunks and branches may show definite direction, to which the eyes can cling, and the whole of a tree or bush may often present a fairly comprehensible sphere or cone shape. Also there may be an over-all texture of leafiness and greenness, but there is much in the landscape that the eyes are simply unable to grasp. And only to the extent to which the confused panorama can be seen as a configuration of clear-cut directions, sizes, geometric shapes, colours, can it be said that it is actually perceived.

[19] Kurt Koffka "Principles of Gestalt Psychology., New York, 1935

The brain processes that make this articulation possible are unknown. We may assume that in response to perceptual qualities, more or less clearly indicated in the raw material of the stimulus, corresponding patterns of simple structure arise in the cortical field of vision. But for the time being, this is pure theory, inferred from what is observed in experience."

Thus gestalt psychologists attach decisive importance for perception to the simple geometrical form or the simple schema resembling geometry. Only those phenomena in our environment which we can perceive in this geometrically simplifying way, which we can bring into agreement with our visual cortex's simple perceptual images as described by Arnheim, can actually count as recognized, as perceived. After this, our tendency towards the simple structure decides whether we understand phenomena in our environment as two or three-dimensional. This is true in the case of perspectivally geometrical forms and also in some other cases.

However, at this point it is necessary to point out that this law in this form fails to explain the spatial effect of most living forms, or does so only with difficulty. I am ignoring one-eyed or two-eyed visual perception here.

As a result of many experiments gestalt psychologists have explained in individual cases, or tried to explain, why we see forms in the same way in which we can experience them by touch, and not as they appear on the retina; which laws govern the way we group and recognize objects; why we experience our surroundings in perceptual units and not, like the freshly operated blind man, as a succession of noises and flashes of brightness. We see trees, hills, houses or the sea and the clouds in the sky, individual people and crowds. We experience these units in the unity of space and time. We organize them in accordance with the principles of similarity, nearness and simplicity. Similar shapes, colours and similar spatial arrangement and, above all, similar movement, combine to form units and separate themselves from less similar or more distant shapes, colours and movements. They explained why shapes and sizes are always perceived as the same; they formulated the law of constancy of shape and size. It is therefore no exaggeration to say that up to 1960 almost everything of importance – and which is frequently also correct – in the psychology of perception came from the gestalt psychologists. Francis Crick[20] also begins his essentially neurophysiological discussions about the brain – in particular about visual perception – with important references to gestalt psychology. He does not, however, unite both points of view. This is certainly not possible at the moment.

I do not want here to repeat the results of gestalt psychology, but to point out instead an extremely important, and so far neglected, area. A fine artist, concerned with expression and pictorial communication is ideally qualified to explore it through much practical experience. Such an artist, after a long life, possesses a huge potential of phenomenological knowledge, but seldom expresses it in words – though there are a few exceptions to this rule: Dürer, Leonardo da Vinci, Alberti or Ghiberti. As a sculptor, I have been concerned with such questions throughout my life and so should like to set down my thoughts on this problem. I would like to ask you to bear in mind the fact that I am neither a doctor, a psychologist nor a neurophysiologist and that I therefore do not always use the standard language associated with these fields. I am unfamiliar with some of the literature that a psychologist or doctor might have read. My thought processes are also different in many respects. But it is because of just these things that I can probably make a valid contribution to this field, as I have worked both theoretically and practically in many areas related to perception for 50 years: I am a professional perceiver and if perception can be said to be an active process – which it is – an artist, who is continually led to new visions of reality by this very process, is able to recognize and answer problems that a psychologist, medical scientist or neurophysiologist does not have to face with such clarity.

Chapter 5
My Question: How Do Forms Convey Content; Are There Visual Categories of Expression?

Apart from the inadequate attempts of Bense's theory of information, gestalt psychologists – and by and large many other sorts of perceptual

[20] Francis Crick "The Astonishing Hypothesis: The Scientific Search for the Soul", op. cit.

psychology – have almost exclusively tried to answer Koffka's famous question: why do we see things as they are and not as they appear on the retina? Gestalt psychologists, like the psychologists and mathematicians working with Lie Transformation Groups, think that in the visual areas of the brain simple geometrical forms are genetically anchored, which symbolize rotation, expansion and parallel displacement. While this is happening on the one hand, on the other hand neurophysiology and cerebral anatomy have made enormous strides in localizing the brain's seeing, hearing, thinking and remembering processes, by discovering chemical, electrical and anatomical states. My question, in view of all this, is: by what paths and methods do things perceived pass on to us information which goes above and beyond substantiating their existence, appearance and position? To my knowledge, this question has never been the subject of scientific research.

It may well be that asking this question can bring together some of the comparatively separate areas of knowledge about the brain and about perception – and these include the development of children's drawing or those suffering from diseases affecting perception. We shall then see how far such thought processes lead finally to consciousness of the self.

If consciousness of the self has something to do with the declaration memory – in the visual area I call this the reproduction memory – then our deliberations will also help to improve our understanding of our consciousness. In my opinion animals have a much smaller declaration memory, while the difference between human and animal recognition memory, short-term and ultra short-term memories are perhaps not very great.

I am asking why every thing has its own expression: a tree does not just tell us that it is there; it also shows us that it has always had to grow against the prevailing wind, or that it has stretched out its branches to the light. A bud shows us that it will open, a leaf that it will unfurl. A human face sends a message telling us that the owner is gentle or choleric, lets us know whether to expect malice or reliability, passion or calm. A person walks energetically or weakly. There is no end to the possible examples. Hardly any form in our – natural or created – environment fails to send us a clear, or at least a vague, message.

The ability to read such messages is essential for our survival. We have to be able to distinguish between what is threatening and what is lovable or desirable. Which are the categories of form, movement and colour which pass such information on to us; which, at a glance, signalize to us essential elements of their fate, both past and future, and thus of their character and being – in other words, let us understand their expression. I am discussing all this only in relation to the form in which things are seen. I am not going to speak about colour or actual movement, about changes relating to place and time.

And this, more or less alone, is the question to be answered in these pages. As we shall see, it is of such central importance for recognition, that answering it will lead to important conclusions about the function of perceptual thinking and will also help to throw a new light on the old question: why do we perceive things as they are?

I am not denying that questions of expression have also been discussed in connection with research into actual movement, but not in the systematic way and, above all, not with the significance that I shall now attach to them.

The human form and its facial features are not alone in having their own expression; every phenomenon in our environment tells us something about its fate and its being. The picture (59) of the two similar pyknic head types with completely different characters shows what subtle conclusions we can draw concerning the intellectual and psychical structure of our fellow beings simply by looking at their features. Without ever having spoken to them we can see that their attitudes to life are completely different and that they certainly judge the same event very differently. What enables us to reach such differentiated conclusions by looking at their form? Can we put it down to experience? If so, we have to conclude that our brains store an enormous amount of visual memory data which we can call up as needed – and that they can do it from our earliest childhood.

Or are there perhaps basic criteria, general form categories, which we can apply to each case as it arises and thus judge it individually? That would certainly be a much more economic principle. Let us begin with the first deliberations.

Look at a ceramic vessel in the geometrical form of a cylinder (61). Now compare it with the three variants whose sides are curved (61 a, b, c), a little differently in each case. Of these four objects, the geometrical cylinder is the least expressive. All it tells us is its geometrical form and its material. It seems neutral, has no real expression, lacks vitality. But that is just one aspect; it is also a low-grade unit. What does that mean?

All forms which differ from others are perceptual units; thus the geometrical cylinder is a unit, but there is a limit to the definition of its position in space. We can turn it through 180 degrees, but it will still look just the same. Its form stays the same if we lay it down; it simply becomes a ceramic pipe. If it were cut into rings, it would not matter in what order we reassembled them. The geometrical unit does not determine the order of its individual units. Nor is the length of this cylinder defined by its form. It could be lengthened or shortened at will, in accordance with the purpose for which it is intended. It could be any length at all – ranging from a flat ashtray to a pipeline.

Thus the geometrical cylinder is a low-grade unit, which tells us nothing beyond its appearance and its material and so has little expression. Unit and expression are interdependent, are conditional on each other, as we shall see.

If we compare the cylinder with the vessel in 61 a, we see that the latter is far more expressive and seems more vital, to expand into the silence, as the Chinese say. Something inside it could be pushing out its sides. We recognize a volume which is vital, breathing, expanding downwards and outwards. This expansive power seems to be pushing against the tension of the vessel's skin. So we are not only looking at a geometrically more complicated form than the first vessel (61), we can also feel that this form may have come about through pressure or intentional expansion. The form's vitality apparently depends on the opposing forces of expansion and cohesion.

The walls of the vessel are under slight tension, that is to say, a curvature which starts at the top becomes stronger under increasing pressure as it proceeds downwards, and finally returns, at the stable foot of the vessel, to the cylinder's original diameter. In accordance with this uneven curvature, we follow the vessel's apparent expansion downwards and outwards[21], towards the largest volume and strongest degree of curvature. The visible expansion accounts for the vessel's vitality. But that is only one aspect.

The vessel has increased not only its expressiveness, but also the degree of its unity. We can no longer turn the vessel through 180 degrees without changing the picture. It is possible, but the expression is clearly different (61 b). The unsymmetrical curvature prevents our turning it through 90 degrees. A horizontal position would just look as if a vertical form had fallen over, and not – as in the case of the geometrical cylinder – like a ceramic pipe. If we imagine the vessel made of horizontal rings, these would, because of their gradually increasing and decreasing curvature from top to bottom, no longer be interchangeable. The vessel's individual sections are fixed by the increasing and decreasing curvature. The vessel's expressive force leads simultaneously to a greater degree of unity.

If we turn the vessel upside down (61 b) we no longer see a downward expansion, the possible result of filling it with liquid, but an upward expansion, which we might interpret as the rising growth force of maturing fruit (63). For the reasons already stated, this vessel is also a higher grade unit.

While in the cases of figures 61 a and b the sides are unevenly curved in steps or gradually, the sides of 61 c are evenly curved. A vertical section through this vessel would result in two vertical circular segments. Here, too, in comparison with the geometrical cylinder we experience an expansion of the volume, but it is not directed upward or downward, has no relation to gravity and thus has less expressive force and seems less vital than examples b) and c). The information it offers us is less exact, and so the degree of its unity is less than that of examples b) and c) but greater than that of example a). We can turn it through 180 degrees without making any difference. It would not be impossible to lay it horizontally, but despite its symmetry, the curvature makes this impractical. If we imagine the vessel to be composed of rings, they are no longer interchangeable in any order,

[21] In order to interrupt the posing of the problem as little as possible, I here ask the readers to form their own judgement by looking at the illustrations. Later all observations will be subject to controls.

but only with analogous rings from the other side. The form does not clearly limit the length, but there are clear limits to its possible alteration, because the circular segmental curvature of the form, unlike the geometrical cylinder with its parallel sides, is not infinite[22].

Thus when a homogeneous geometrical cylinder, which in any event is vertically or horizontally regulated by its longer axis, is deformed by a regulated expansion, this expansion can also be distinguished in the form thus attained. At the same time this produces a unit of a higher degree. But before we draw conclusions which go even further, let us look at even and uneven curvature in the case of the epitome of curved geometrical forms: the sphere.

The sphere is the most evenly formed of all the three-dimensional images imaginable. All points on its surface can be attained with the same radius. The surface's degree of curvature is always the same, so that all radial sections and all areas of the surface are interchangeable. This total uniformity equals total lack of direction. The geometrical cylinder lacks differentiation between top and bottom, but it is regulated in its longer axis. This can be seen if it is laid horizontally. Because of its shape, it looks "peculiar" if it is tilted. But a sphere can be turned in any direction in space without changing its appearance. All parts of its surface are interchangeable with each other. To that extent the sphere is a unit of the lowest grade.

On the other hand, the sphere's proportions are absolutely unalterable. The enormous simplicity of its form and the inalterability of its proportions have always fascinated artists and continue to do so. They thought they had found, without any doubt, ideal proportions in the degree of its curvature. At the end of the eighteenth century the sphere was rediscovered as the – alleged – most perfect form. This came about as an understandable reaction to the infinite freedom of late rococo's unevenly curved and frayed forms. In contrast to the total deformation of all simple geometrical forms in the middle of the eighteenth century, they now discovered the unsurpassedly simple form whose compelling geometry prohibited any alteration.

In 1784 Vaudoyer, for instance, designed a spherical house, an imposing residence (65), probably for a ruler. He designed it as an ideal. New forms were sought, without thought of practicality. It is legitimate to ignore reality, when in search of new ways and possibilities. But was this really a new way? Evenly curved, without top or bottom, the sphere can roll away without direction if not firmly held. In this case it is laid over an axis on a circular architrave, supported by columns. It is suspended like a globe. But the sphere, without direction or relation to gravity, has no visible connection to its peristyle. It floats in the architrave, but fails to form a unit with it. It could rotate, or, at any moment, float off like a balloon. A vast spherical monument was designed for Newton's grave, with a diameter of more than 100 metres (66). However a sphere is placed, its lack of direction prevents its combining visually with other forms or even with the ground it stands on.

This changes when the sphere is cut into, when the result is a segment of a sphere. While Vaudoyer's spherical house floats completely without direction in its peristyle, Fuller's segment, resting on its flat surface, directed upwards, resembles a captive balloon (67). Through the contrast between its cut-off base and spherical roundness, the segment of a sphere seems to be directed towards the vault. Fuller did not make very much of this in the form of the pavilion illustrated here. Now let us consider Pier Luigi Nervi's little sport palace, built in Rome in 1960 (68). This concrete dish is a – much smaller – segment of a very large sphere, also apparently directed upwards towards its vault, like an inflated tent. Nervi developed his design from these givens: the segment of a sphere rests on vector-like, sloping, y-shaped concrete supports. Because of the optically upward movement of the segment of the sphere, we see these as tent ropes, holding down the inflated cloth. This impression is strengthened by the wavy border, which also stiffens the thin edge of the dish.

The tent, pinned to the ground by the y-supports, seems to swell out over the actual boxing arena, thereby making a historical motif to a theme of the architecture. In ancient Rome, when it rained, marines in the topmost gallery protected the arenas by unrolling a tent over the seats.

[22] Jürgen Weber "Gestalt, Bewegung, Farbe" 3rd edition. Braunschweig/Berlin, 1975/1984

The body of the Hellenistic beaker from Athens also consists of a sphere with its top cut off and replaced by a cylinder (69), so that the entire vessel bellies out downwards.

However, this direction is even more obvious when the even curvature of the segment of the sphere is gradually unevenly deformed, as in the case of a drop or an egg.

A drop of water, leaping up and falling back in a fountain, alters its shape from drop to sphere and back to drop. As long as it is moving, the main mass of its volume pushes forward, forming a blunt end, while the rear end tapers off. It is a pure, immobile, totally symmetrical sphere only in the instant when it reaches the top of its path, at the moment of indecision between rising and falling. A drop is a sphere deformed by movement. Its longitudinal section is only axially symmetrical. It bears witness to the movement from which it developed. The drop seems to rise or fall in the direction of its blunt end. In our eyes the blunt end is the front and the sharp end, the back.

The drop's surface areas are interchangeable only within its circular cross-sections. So it is not only true that the drop developed from the sphere by means of movement – and expresses this for us – but it also developed into a unit of a higher grade than the sphere.

Let us examine a ceramic vessel shaped more or less like a drop. It is cut off at the bottom so it can stand, so it is also a segment. We experience the directional pull of the expansion of the volume and of the increased curvature of the surface even more clearly than in the case of the cut-off sphere. The swelling form (70), directed downwards, seems to express the weight of the contents. If we reverse the direction of the vaulting, so that the strongest swelling looks upward, the form again appears to rise or to swell upward like a fruit (71) and like the two prehistoric Egyptian pots (64, 72). The expression of such a vessel contains a contradiction, seems to be the result of movement and counter-movement, or force and counter-force. The volume seems to expand, while the vessel's skin extends itself as a counter-force around the active volume, thus making the dynamic process visible. We also experience the form in a contradictory manner: on the one hand we have the feeling of an expanding volume, on the other hand the tense curve of the vessel's skin seems to be working against an alteration of the form. We feel that its exactness must be final. It is this contradiction, this battle between a form-altering and form-retaining principle, that gives the vessel its vitality. It is here we find the basic difference to the inalterability of the sphere. Since it has no direction, the finality of its curve does not contradict any pulsating, growing volume, as do the drop or swelling–forms – as their expression leads us to call them. The inalterability of its proportions results in dull, expressionless rigidity.

Another factor in this is gravity, which also plays a part in the way we experience form. Upward expansion of the volume seems more dynamic, more forceful and more elastic than the opposite movement downward. This is because we sympathize with the force working against gravity, conquering it by means of rising volume. A form swelling downward seems less active, but instead fuller and heavier. Two similar forms appear quite different, their expressive force quite contradictory, because one is, and one is not, "resisting" the pull of gravity.

In order to test how many observers agree with this conclusion, in 1994, during their first lecture at our university, we set the first-year architecture students – straight from school – a questionnaire showing a ceramic vessel swelling upward and another swelling downward, together with the following questions (Questionnaire I):

Do you see the volume of the vessel on the right/left as pushing upward?
Or do you see it as heavy and directed downward?
Or do you not see any direction at all?

Simultaneously we projected transparencies showing the two vessels onto the wall, to compensate for their inadequate reproduction on the photocopied questionnaires. We added nothing to the questions on the paper, except to ask the students to look at the pictures carefully, to work quietly and not to look at their neighbours' papers. I have to stress here that attending my lectures is voluntary. They are just for practice; no examinations are taken on them. The students' final results are graded solely on their sculptures. The undergraduates know this. Occasionally students who have attended hardly any lectures at all get good final results. So many corrections are made in the art classes that the students can undertake the tasks easily.

I

Do you see the volume of the vessel on the left as pushing upward,  97,4 %

or do you see it as heavy and directed downward? ☐

Or do you not see any direction at all? ☐

Do you see the volume of the vessel on the right as pushing upward, ☐

or do you see it as heavy and directed downward? 97,0 %

Or do you not see any direction at all? ☐

II

21.10.1994

Does the line to the left show movement?

- ☐ from right to left
- 14,6 % from left to right or
- 83,0 % in no direction at all

Does the line to the right show movement?

- from right to left 65,5 %
- from left to right or 33,9 %
- in no direction at all ☐

Does the line to the left show movement?

- 16,3 % from right to left
- 83,1 % from left to right or
- ☐ in no direction at all

It goes without saying that no names were written on the questionnaires. Filling them out was completely voluntary. Most of the undergraduates had fun; they were, in other words, motivated. It should also be noted that some of the questionnaires handed in were not completed. After five minutes we established that all the students had finished without saying anything, just using signs. We then collected the papers. 97.64% of 169 students saw upward movement in figure A. The remaining 2.5% saw either a downward movement or none at all. At this point I would like to repeat that my first lecture to undergraduate students in their first semester has always been their very first lecture at our university.

97.05% saw a downward moving heavy volume in figure B, 2.95% saw an upward movement. It is therefore safe to say that unevenly curved vaulting is seen as moving in the direction of the strongest curvature and volume expansion. That was the first questionnaire on this subject. Later repetitions confirmed these results.

In this connection we then found it interesting to check whether we would obtain the same result in the case of an unevenly curved line – a streamline. So on the same day we distributed a second questionnaire (Questionnaire II) with a segment of a circle and an unevenly curved streamline in two positions, from left to right and from right to left. We accompanied each line with the same three questions:

Does the line show movement:
from right to left
from left to right
in no direction at all?

For the segment of the circle, A, 83.04% said "no direction at all", but 14.5% said "left to right" i.e. the direction in which we read.

For the unevenly curved line starting flat on the right and finishing strongly curved on the left, B, 65.5% decided "right to left" i.e. against the usual reading direction. However, 33.9% read the line in the reading direction and said "left to right", from the strong curve to the flat part. For the line curving in the opposite direction, developing in the reading direction from flat to strongly curved, C, 73.1% said "left to right", but there were still 26.3% saying "right to left", against the reading direction. So the result is less conclusive than in the case of the unevenly curved vaulting. A percentage of between 26 and 34% read the line from the strong curve to the flat part. Since we must surely take into account the fact that we are used to reading from left to right, we can probably estimate that some 70% of the 171 persons tested read the line from its flat end to its strong curve, while 30% read it in the opposite direction. To make a basic point about all our questionnaires, I would like to say that they can be used and therefore checked by anyone. The only condition is that there is a large enough number of people being surveyed, otherwise the results will depend too strongly on the chance composition of those questioned.

In conclusion we can say that we read unevenly curved vaulting in the direction of its volume expansion and strongest curvature, while the unevenly curved line usually also seems to be directed from its flat end to its strongest curvature, but that it is also possible to read it in the opposite direction. In nature we also find both unevenly curved vaulting and unevenly curved lines, so we shall have to return to this result.

There is no doubt that we also judge natural growth processes in accordance with this law. The vital force of the lily blossom (73), developing and opening forward, is primarily made visible by the three-dimensional, increasing vaulting of its petals, developing and opening forward and outward. Similarly, the growth of the pepper shown here (74), seeming almost explosive, is made clear by the increasing vaulting of its separate divisions. However, it is certain that in both cases the figure of the rosette plays a role. We shall have to return to this later.

In the case of an apple hanging from a branch (75), we read the main direction of growth from the blossom to the stalk, i.e. from the flat to the strong vaulting, though we also recognize a secondary direction toward the little, rosette-like divided vaulting around the apple blossom (76). In nature there are often two directions towards the main vaulting and towards the secondary vaulting. We also see the aubergine as growing towards the direction of the volume expansion and the strongest curvature (63). The granite rock (77) seems to be swinging upwards like a living creature, from bottom right to top left, because this movement is suggested by the double process from flat to strong vaulting, combined with its diagonal direction. This impression is strengthened by the enlargement from the lower to the upper part of the form. (Test results on this point will be presented below.)

The vessel with handles (79) consists of a large body, whose strongest curvature points upward, while it bears a flatter, smaller body, whose increasing curvature points downward. The two separate parts of the whole volume appear to be pushed together by pressure, in accordance with their opposing directions. The larger form appears to be bearing the weight of the smaller.

The second vessel (78) is quite different, consisting as it does of a succession of a smaller and larger form, both with vaulting pointing upwards. We read the form as starting to rise twice, developing apart. At the top it appears to open like a blossom. Apart from this, the similarly directed curvatures of the two forms are in accordance with an increase in the volume from the bottom to the top, so that we see the whole volume rising upwards with increasing force. The opposite is true of the vessel illustrated in (80), where the two parts of the form seem to be pushing against each other, similar in principle, to the vessel shown in (79).

Our results to date can help us a little to explain the opposing expressions of the two heads (59) we noticed at the beginning. While in the case of the man on the right upward vaulting determines his entire face, from his chin and mouth to his cheeks and cheekbones, thus giving it a vital, gravity-opposing expression, on the left-hand face all the vaulting, beginning with the cheekbones and cheeks, down to the mouth and jaw, is directed downward, so that it has a resigned, sceptical expression. The differing strength of the vaulting also plays a part: the sailor's downward directed vaulting is very flat, while the ship owner's corresponding features spring forward more strongly, so that – quite apart from the up or downward direction – we see a more vital or weaker expansion of the volume. But we must deal with these problems separately, in connection with the rosette. Beside this whole opposing picture there are also some individual points of agreement. In both men, the neck and palate area spring upward, while the vaulting of their chins is directed downward, so that as a result of the similar way they are holding their heads, both of them seem to be pressing the chin against the neck and palate.

So we can say, for a start, that the up and downwardly strongly or weakly curved vaulting in the same direction or opposing each other, goes some way towards explaining the expressions on both faces – but, of course, not all the way.

I would like to stress here that I do not want to teach the reader how to "see", but rather that we are looking for categories that the visual areas of the brain use to form judgements.

Let us recapitulate what we have said so far: We have compared the unevenly curved vaulting, surface or line with the even curvature of the sphere, the circle or the uncurved line. We understand the departures from these simple geometrical forms as the result of a process, a past movement. So, for us, the uneven curvature expresses a directional movement and, simultaneously, gives the form in question a higher degree of perceptual unity. This is what I describe as the metamorphosis of geometry.

Simple constellations in the cerebral cortex in the visual areas match such simple geometrical forms as circle or rectangle. For the moment that is our working hypothesis, which is supported by the development of a baby's perceptual abilities, by children's early drawings and the phenomena of visual agnosia and Alzheimer's disease. This has already been claimed by the gestalt psychologists and, rather differently, by Hoffman/Dodwell and others. But I now claim that we compare nature's more complicated forms – in our initial experiment the drop or streamline – with these simple Euclidean forms. In our visual cortex we understand the deviations from possible geometrical prototypes as the result of a past or present movement or process, in any case of an alteration – and also as spatial. At the moment this claim may sound rather daring, but could become plausible, if we could prove that this theory could be used to provide the most simple explanation of vastly different perceptual phenomena.

We have shown that a segment of a sphere resting on its cut-off surface, seems to follow the direction of its vaulting. The same is true of the section of a circle, sitting on a base line (81 a). But if the segment of the circle stays open – as long as it is more than a semicircle – the observer feels that it needs completing (81 b). For example, that is the basis for the vital and tense effect of the arches in the mosque in Cordoba. They consist of roughly two-third circles (82) which we try to imagine completed, probably because we want to make them agree with the circular forms in our perceptual images. However, because in each case two arches meet symmetrically on the capital of a column, the simple form of the symmetrical capital and the contrary direction of the circumference of the circles contradict the observer's compulsion to complete the circle. The "silent drama" of these arches arises from this unsolved conflict between circles wanting to close but being hindered from doing so by others, this game of movement and counter-movement. For the same reasons a similar vitality is attained by neo-classical buildings' windows and doors which slightly exceed a semicircle. The slightly more than semicircular segment is prevented from closing by the rectangle (83). Moreover, the fact that the closed circle among incomplete circles is hard to find but that the incomplete circle among closed circles is found easily (Treisman) is explained thus: the unclosed circle differs from our perceptual concept of a circle and is thus noticed more readily, this is not true of the complete circle (Crick) (81 c).

Chapter 6
The Rosette

The phenomenon of the rosette, turned outward or inward, is also part of this complex (Questionnaire III)[23]. If we assemble a complete circle absolutely symmetrically from little segments of circles, we get a dynamic form which is clearly different from the simple circle. Unlike the circle, the rosette seems to want to expand from its centre outwards to the little segments of circles. It appears as a dynamic figure in front of a background.

A counter-experiment shows to what extent the circle's original shape is activated by the symmetrically organized semicircles. If we set the semicircles around the inside of the original circle (Questionnaire III), relations between base and figure change. We no longer see an expanding circle. We see surroundings trying to close in towards the centre (cf. Arnheim: "Kunst und Sehen", p. 194). Both figures seem to grow in the direction of their semicircles and not in the direction of the acute angles. So they demonstrate a continuous change between active and passive zones. This impression is not hindered by the central symmetry. The questionnaire we distributed among 168 first-semester architecture students in their second week was answered, in the case of both figures, in accordance with this result by 74% to 78% of the students. The remaining 22% – 26% was divided between three answers and so can be set aside.

Gothic architecture made frequent use of such figures to lend vitality to circular windows (84, 85, 87). A symmetrical rosette is pushing against the circular edge of the window opening. The impression of expansion is hardly any less if the rosette's foils are fastened to the centre by pillars. The dynamic of this figure can be increased by the asymmetry of a fan window (84, 4). This could also be the explanation for the radiant effect of the peacock's tail (86). Since the primary purpose of the rosette segments in the peacock's tail is presumably to impress the peahen, we are justified in concluding that similar principles apply to the evolution of human visual perception, which goes way back in time. The rosette could have something to do with propagation not only in the bird world (peacock, pheasant, capercaillie (86, 89) etc.) but also in the insect world. Nor must we forget that most blossoms are variants of rosettes (91, 92, 93). A woman's breasts are also the segment of a rosette.

The arches over portals which show Islamic influence in the south of France are, from the end of the twelfth century onward, directed more strongly outward or inward in accordance with the outwardly or inwardly directed rosette. In the case of the portal of Le Dorat, little segments of circles increase the impression of the archivolts' wavelike movement, spreading across the surface (90). In the portal of Ganagoble (88) we find segments of circles directed inward on every second arch, which seem to push against the portal. So, on the one hand, because of the gothic archivolts expanding outward, we experience a wavelike outward movement across the surface, on the other hand we can understand the walls with their stonework as a figure closing around the portal's opening. In this way the stonework formally expresses the door's double function: opening and closing.

[23] Rudolf Arnheim "Kunst und Sehen" (Art and Visual Perception), op. cit.

III

Number of questionnaires handed in = 168

Please compare both depictions of a rosette, then read all the possible answers before ticking two answers for each question.

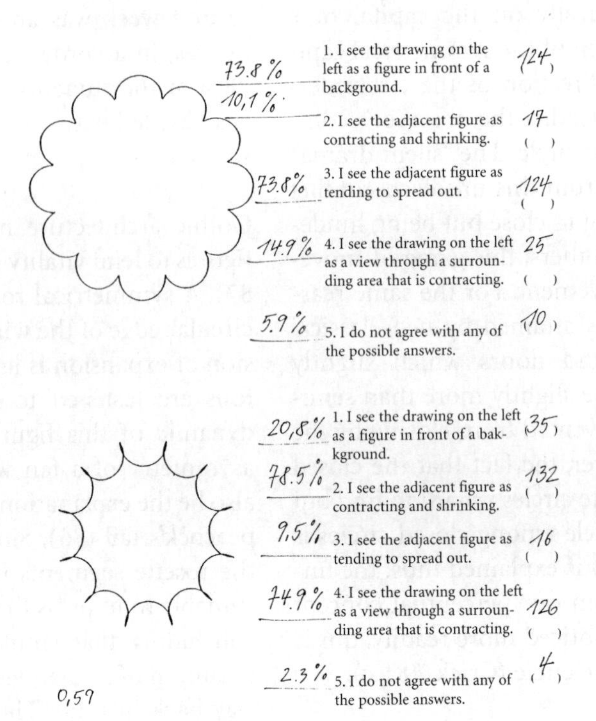

IV

Please read through all the possible answers, look at the figures, possibly holding them at arm's length, and tick just one of the five answers for each figure.

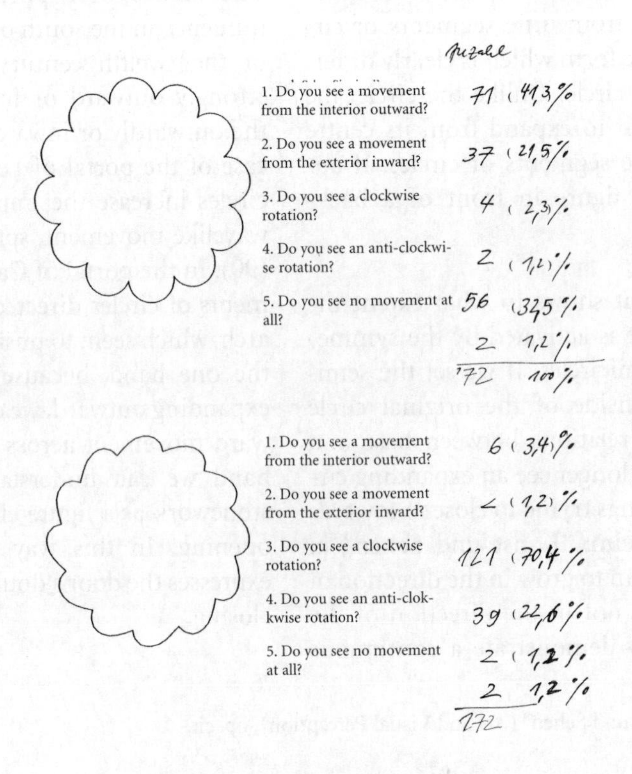

Rosettes do not have to be made of segments of circles. The rosette from the church of Assunta in Altamura (87), from the beginning of the fourteenth century, consists of arches made of mules' backs which, spreading over two radii, penetrate each other. Rosette windows appear in every possible variation in gothic architecture.

If we alter the segments of the circle to form uneven, bent, streamlined figures (Questionnaire IV), they start to rotate. 70.4% of 172 test persons see the rosette turning in the direction of its strong curvature, i.e. clockwise. Admittedly here, too, 22.6% see it turning from right to left, i.e. from the strong to the weak curvature. The unevenly curved streamlined rosette also occurs naturally; for instance, white oleander blossom. In any case, 93% see rotation. It is interesting that in comparing the dynamic streamlined rosette with the simple rosette, 32.5% see no movement in the simple rosette and only 41.3% a movement from the interior outward – though even this is double the number who experience the opposite – while in comparing the outward or inward directed normal rosette, between 74% and 78% of 168 test persons see a spreading or contracting movement of the little semicircles. This shows once again how strongly we are influenced by comparison with a less dynamic figure. However, we understand the unevenly curved limits of a surface as the distortion by movement of a circular surface.

These examples have shown us that – unlike the case of the peacock – a rosette need not necessarily consist of true segments of a circle, but can also be made up of distorted segments – as in most cases in nature. In the three-dimensional area this even creates an intensification of its expansive effect. In the case of vessels consisting not of circular, but of rosette-like cross-sections (94), the individual horizontal segments of these rosettes constantly change their form in vertical succession. In the case of the vessel illustrated here, the rosette segments begin at the bottom deep and like semicircles, in the area of the strongest vaulting of the whole rising form they change to completely flat arches, and finally become deeper again; that is, they alternate between rather semicircular and completely flat vaulting, only to finish up resembling parabolas at the top of the vessel. It becomes apparent that it is exactly this change in the rosette segments that is an intensification of their expansive, vital and cohesive effect. In the other vessel (95) the rosette segments were, as far as possible, kept the same from top to bottom, which is why the negative elements between the segments, the foils in the rosette's cross-section, are almost equally deep from top to bottom. The effect of this is to divide the vessel into several vertical sections, so that we see its body less as a unified, breathing, expansive volume. Rather it seems to be constructed of vertical individual parts which, although vital, seem to be assembled rigidly, despite their very organic and plastically experienced uneven curvature. An absolute precondition for vitality of expression is a change in the geometric form or a distortion. It is not only necessary for the vaulting of the vessel and the rosette to change, the negative forms, arising at the meeting points of the rosette's parts, must change gradually from flat to deep and back again, if they are to make a vital, breathing impression that is also coherent horizontally.

The starting point for a rosette is not necessarily a circle. The exceptionally expansive volume of Dietzenhofer's Trinity Church in Kappel (1685 – 1711) comes about solely because of its rosette ground plan (96), which in this case is based not on a circle but an equilateral triangle. Its dynamic appearance is considerably increased by the alternating small and large segments of circles (96, 97). In the case of the Ganagoble portal, dating from the end of the twelfth century, we found not only the pointed Gothic arch as a basic form for the rosette, but even the rectangle of the door opening (88).

The rosette's central symmetry probably also corresponds to a class of form in our visual cortex, perhaps a variant of the radius "orbits" or circular forms of the Lie Group (50). That is why many people find particularly beautiful not only the rosette windows in Gothic churches but also the radiolarians, jellyfish and coral shapes first drawn by Haeckel (48) in the middle of the nineteenth century. Probably both the centrally organized rosette and Haeckel's centrally organized natural radiating forms correspond to our visual cortex's point symmetrical organization system. Interestingly Haeckel even called these centrally organized natural forms "nature's art forms"[24]. Artists of all eras have constructed anew nature's pictures and the world around us, distorted by the geomet-

[24] Ernst Haeckel "Kunst-Formen der Natur". Leipzig/Vienna, 1904

rical preconditions of our visual cortex. Why was point and axial symmetry almost always and on every continent such an important artistic and ornamental organization principle? We shall return to this question.

As I have already said, the development of infants' drawing supports the idea that our visual cortex also contains point symmetrical organization systems. At the end of a child's fourth year, shortly after it has developed the circle, the ladder (screen-system) and the rectangle, it adds radials to the outside or inside of the circle (34, 37) and then develops a tadpole.

I assume that the ability to develop geometrical forms from basic processes is genetically programmed. On the other hand, comparing environmental phenomena with these basic Euclidean forms and constructing pictures that can be remembered from them is something that comes with experience. This certainly has something to do with the self organization of the brain as described by Singer.

It is from this point symmetrical form that the child develops its first human schema, the tadpole – even though the human form is by no means point symmetrically constructed (35). But the point symmetrical radial image is probably – beside the ladder or screen system – the geometrical system that develops earliest and on which various individual images can be assembled to a whole image. Just as the screen system later helps children to organize their pictures (41, 43) – as we shall see later – it also develops pictorial organization from the point symmetrical radial figure, for instance, people around a table or at a circus (98). Of course, children have no way of drawing complicated perspective intersections, in order to produce a correct spatial depiction of people around a table or at a circus – nobody is denying that – but, nevertheless, it is not self-evident that they are constructing the spatial connection from the same point symmetrical system that acted as the point of departure for developing the first human schema and, occasionally, the animal schemata (35, 38).

In conclusion I should like to point out, that we experience many natural forms as dynamic because they are constructed on the principle of the variform spatial rosette. This applies not just to the flower forms – which give the form its name – many vaulting processes of plant, animal and human forms are also based on the spatial-plastic effect of a varied rosette system. Examples of this are the pepper (74), the apple (75, 76), the horses (100, 102), whole treetops and their parts (103) and cumulus clouds (113). The dynamic effect of my "Brunswick Venus" (101) and of Michelangelo's "Night" (99) – as inward directed rosettes – are also based on this principle. With nearly all these examples it is the change from more to less vaulted segments and the opposing degrees of curvature that increase their expansive dynamic – in a similar manner to the Dietzenhofer chapel, where the little rosette segments of the staircase towers alternate with the large segments of the interior of the church and its ambulatory.

Just as the streamlined rosette (Questionnaire IV) is understood as an expansive rotational movement when compared with the semicircular rosette, so all natural rosettes – both flat and spatial – are understood in their particular direction by comparison with rosettes which are spherical or resemble semicircles.

Many eighteenth-century interiors with their variously large and variously forward-extended galleries, columns and profiles (110) can provide examples of architectural processes employing in or outwardly directed rosette-like shapes. Other epochs provide whole series of rooms, ranging from, for example, St Vitale in Ravenna (105) to the Hagia Sophia (106, 107) and the dome of the cathedral in Florence (108).

There are far more natural processes resembling rosettes than we realize, but such systems are often easier to recognize in art than in their natural prototypes. We shall be looking at the individual examples.

Chapter 7
Contraction and Expansion

This chapter is somewhat difficult to read, because of the large amount of numbers and questionnaires in it. It may even appear to be a little boring. I would, however, still like to ask the reader to work through this chapter, as this will lead to a better understanding of the work as a whole.

Summing up – a simplified version of our thoughts so far – we can say that we compare unevenly curved lines, surface limitations or bodies in our surroundings with their corresponding simple geometrical images in our visual cortex. For example, we compare the slightly curved entasis of the Doric column (109) with a cone or cylinder, the column's cannelures with an inward-turned rosette, whereas we compare drops with circles or spheres. We recognize variations on the geometrical structures as a result of processes which, in the three-dimensional area, are almost always read in the direction of the strongest curvature, but for which, in the two-dimensional area, there is only a preferred direction (74%/26%), towards the strongest curvature.

In this we proceed from the hypothesis that such Euclidean figures as the circle, rectangle, ladder or screen-system and the point symmetrical radial figure are inventions of our visual cortex. We compare these with our environment's natural phenomena in order to understand them both in part and in whole. Again and again we were confronted with the phenomenon that the more complicated form arising from a deformation of a simple geometrical form not only makes this process understandable again; it also promotes it to a unit of a higher order. It becomes more definite and it becomes less and less possible to exchange the succession of the elements forming it – for instance, the individual sections or the surface areas, it becomes more "binding".

So we can say that the images of our environment which correspond exactly to our brain's perceptual concepts pass on practically no information over and above their own geometrical identity. It is left to deformation and deviation from our geometrical ideas to tell us something about our surroundings. *This is the second basic law of perception.* The first basic law of perception was determined by the gestalt psychologists. We always see as simply as we can, under the prevailing conditions. Attneave[25] developed similar ideas in 1954, though only in connection with the gestalt psychologists' "good form" and their signification concept.

It is possible to counter this theory by saying that the deformations mentioned – for instance, the conversion of a sphere into a drop – actually correspond to physical forces and that we judge them not through comparison with our visual cortex's simple geometrical images, but through experience. This would demand an enormously powerful visual memory, a rather uneconomical principle, and, apart from this, I think this argument would only lead us back to the old question: Which came first, the chicken or the egg? Evolution has given the brain the ability to develop geometrical perceptual images with which, as a rule, actual physical processes can be understood.

Essentially the human brain does not mature until some time after birth – some areas take years. Only then is it fully functional. However, that would still be amazingly early for the assumption that our visual judgement relies essentially on experience, for children are apparently able to judge their visual environment more or less correctly from the age of five or six, and in some areas much earlier. Our theory, however, allows the detailed understanding of natural forms on the completion of the simple brain constellations with which natural phenomena can be compared. It is certain that animals' simple perceptual images are present at birth (Robert Fantz's experiments in 1960 with newly hatched chicks).

It is perhaps possible to say that our perceptual images follow a schematism of simplicity. Our geometrical imagination presuppose equal sizes, equal spaces, equal angles – initially probably only right angles – and equal degrees of curvature. Acute-angled triangles are the last to appear in children's drawings, but probably also belong in this category. Because of the analogies I have indicated between perceptual development in the first post-natal weeks and children's early drawings, researchers into the behaviour of babies should perhaps allow themselves to be led by these early drawings – when, in which week, for instance, is the triangle recognized as a geometrical form?

Let us begin again with the "orbits", with the point symmetrical radial figure drawn by the infant immediately before it invents the tadpole (34, 37). At the age of three or so, the infant draws radials around a circle – similar to natural rays – which, evenly spaced, point in the direction of the centre of the circle. Soon afterwards it draws radials which meet there (37). We wanted to know how

[25] F. Attneave, "Some Informational Aspects of Visual Perception", Psychological Review, 1954

A V

Do you find that the radials of figure A

cannot be evaluated	4	2,51%
are stiff and unmoving?	4	2,51%
are rotating slowly?	13	8,17%
are rotating more quickly?	14	8,80%
are rotating and stopping suddenly?	124	77,98%
	159	99,97%

B

Do you find that the radials of figure B

cannot be evaluated	4	2,31%
are stiff and unmoving?	137	86,16%
are rotating slowly?	14	8,80%
are rotating more quickly?	3	1,88%
are rotating and stopping suddenly?	1	0,62%
	159	99,97%

4.11.1994

C V

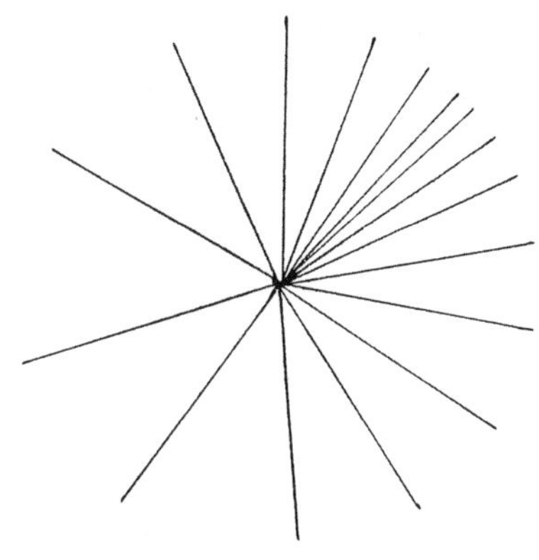

Do you find that the radials of figure C

cannot be evaluated	4	2,51%
are stiff and unmoving?	14	8,80%
are rotating slowly?	108	67,92%
are rotating more quickly?	19	11,94%
are rotating and stopping suddenly?	14	8,80%
	159	99,97%

D

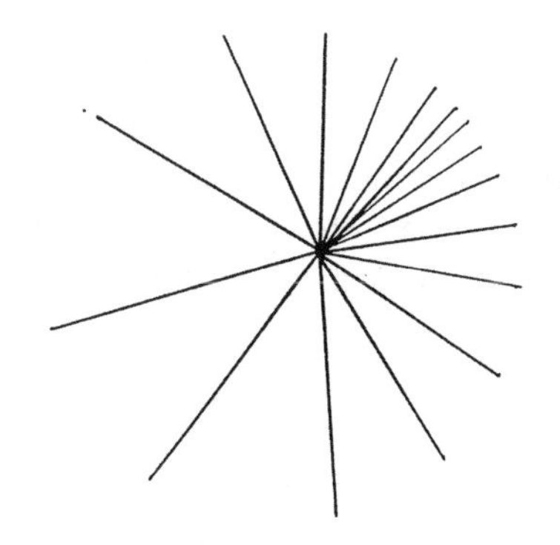

Do you find that the radials of figure D

cannot be evaluated	4	2,51%
are stiff and unmoving?	15	9,43%
are rotating slowly?	26	16,35%
are rotating more quickly?	103	64,77%
are rotating and stopping suddenly?	11	6,91%
	159	99,97%

4. 11. 94

V

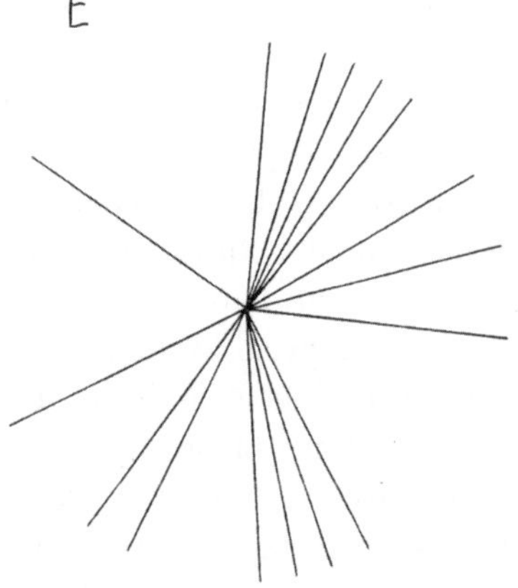

Do you find that the radials of figure [E]

are stiff and unmoving?	12% ☐
are rotating slowly?	0,7% ☐
are rotating more quickly?	2,9% ☐
are rotating together?	2,2% ☐
are rotating separately?	77% ☐
are rotating and stopping suddenly?	7% ☐

adults judge such point symmetrical circular figures with radials evenly and unevenly spaced.

To this end we presented first-semester students during their second lecture with a questionnaire containing five radial figures: A, B, C, D and E (Questionnaire V five radial figures). Only figure "B" is composed of 16 radials of equal length evenly spaced as in the radial "orbit" A (50 A). The radials meet in the centre of the circle. We did not draw its circumference. In children's drawings we find both: radials aiming at a central point but set on the outside of a circle, and radials like these – but always more or less evenly spaced (37). We presented five versions of this basic form. In "C" the spacing between the radials varied from narrow to wide and back again, while in "D" the centre of the imagined circle has been pushed so far to the right that not only the spacing between but also the length of the radials varies. While in the case of the two figures described last the alterations resembled sine curves and were harmonious, the corresponding alteration in the case of "A" ran in a linear manner from close to wide, so that the shortest and closest radials were right up against and in great contrast to the longest and widest ones. Only four alternatives were offered for judging the figures (apart from E).

The 159 questionnaires resulted in 88% of the test persons seeing "B" – radials of equal length and spacing – as "stiff" and "unmoving", while 80% saw "C" – gradual alteration of radial spacing, retaining equal length – as rotating "slowly" (68%) or "fast" (12%). The remaining 20% were divided among the other three possibilities and unusable answers.

Over 80% also saw "D" – alterations not only in the spacing but also the length of the radials – as "rotating", but this time almost 65% said "fast" and only somewhat over 16% "slowly".

It therefore appears that when both the spacing and the length of the radials are altered, the effect of the figure is more dynamic. In any case, in contrast to the simple schema with equal spacing and length, a rotation, movement, is experienced, even though this by no means belongs to the optical impressions of our daily life. The spokes of a stationary, slow or fast-moving wheel do not seem to change their length and spacing; they just go out of focus.

If the changes in spacing and length follow a sine curve, i.e. have no strongly contrasting breaks, we see the rotation as connected and unified. But if the spacing and length changes linearly, so that the shortest radials with the closest spacing are directly next to the longest ones with the widest spacing, rotation is seen as antagonistic, directed against each other. 80% of the test persons described "A" as circling and suddenly stopping. So they saw the increasing speed of the radials at the point of confrontation between close and short and long and wide as directed against each other, and so, as coming to a sudden stop. (We shall be returning to this result with another experiment.)

We must now return once more to "B" with its 12 radials of equal length and spacing. (Questionnaire VI A/B two radial figures) If we turn the figure just a little, so that one of the diagonal pairs crossing at right angles becomes vertical and horizontal, the figure's whole expression changes. On the one hand, if "A" is turned to the horizontal and vertical level, we are immediately aware of this linear cross, in contrast to "B", where neither of the diagonal crosses is particularly noteworthy. On the other hand, the radial figure rotated to the vertical seems more static and fixed than "B", which has no verticals and horizontals.

We checked this on 14 April 2000 by using a questionnaire with only these two radial figures. Of 122 test persons, 94.32% saw "A" as "stiff" and "B" as "rotating" 73.8% or as "moving indefinitely" 23.8%.

In the preceding questionnaire, when contrasted with three variations with variously wide spacing and variously long radials, "B" was seen by 88% of the test persons as "stiff". Now, contrasted with a radial figure rotated to the horizontal-vertical level, it suddenly appears to be moving. What can be the reason for this?

We could point out that, because of gravity, the vertical and the horizontal are – for quite general statical reasons – two particularly important directions. But at a time when nature was untouched by human hand – during the epochs in which our perceptual system was developing – strictly vertical objects could be seen far less frequently than those favouring most other directions – and we are here concerned with seeing. Tree trunks, for instance, are seldom vertical (111, 112, 256). Although even then horizontals played

VI

14. 04. 2000

Please compare figures A and B and tick one answer for each figure.

A

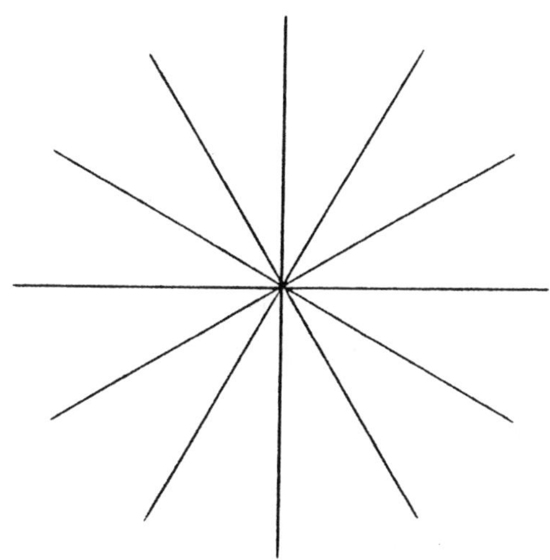

Do you find that the radials of figure A

are stiff and unmoving?	(115)	94,3%
are rotating?	(–)	-
are moving indefinitely?	(7)	5,7%
	122	100%

B

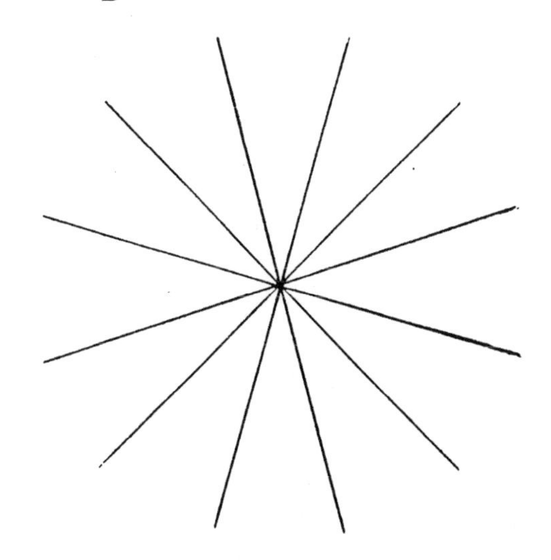

Do you find that the radials of figure B

are stiff and unmoving?	(3)	2,4%
are rotating?	(90)	73,8%
are moving indefinitely?	(29)	23,8%
	122	100%

a certain optical role, it was smaller than some people think. The surface of the sea, from above and below, some parts of sandy deserts, lakes and the course of rivers are horizontal, even if they are often pictured otherwise on the retina. Nevertheless, the horizontal-vertical cross-hairs in our "A" catches our attention first and leads to the stabilizing effect which nearly all observers interpret as stiffer and more immovable.

I answer this question with the assumption that the horizontal-vertical cross in our radial figure "A" consisting of rectangular crosses, corresponds, in contrast to the radial figure "B", with two of our visual cortex's images: the concentric "orbit" figure with regularly spaced radials, which appears in children's drawing shortly before the development of the tadpole, and the rectangular ladder or screen-system, or the rectangular cross, constructed on the horizontal and vertical direction (24, 29, 30, 31 a, 44). We shall have to return to this. Here is just a thought: because, on the one hand, the horizontal/vertical-system, on account of gravity, is so enormously important for what happens in this world, but, on the other hand, appears so seldom exactly on the retina, we have developed a horizontal/vertical image in the form of the ladder or screen-system for analyzing our optical impressions. That is why a picture hanging crookedly on a wall is so unbearable. Coming back to radial figure "A": because the horizontally/vertically directed radial figure corresponds to two of our visual cortex's constellations, we notice the horizontal/vertical cross first and it additionally stabilizes the figure in accordance with the screen-system.

Our thought model – the second basic law of perception – on the basis of which we see deviations from the schematism of simplicity as the result of a process, would explain, in the case of the five different central figures, the varying perception of rotation, whereas we would have to call upon rather complicated interpretations to trace these phenomena back to optical experiences in our daily life. It would be just as difficult to base these different impressions on the law of the simple form.

The following comment on our experiments is necessary: the extremely different results concerning the two radial figures with equal spacings and lengths show how important it is to consider which figures are being compared to each other. The answers are always relative. We are not dealing here with physical or chemical reactions. The difference between the figure with equally spaced radials – it does not matter whether they are set up vertically or not – and the figures with unequal spacings and lengths is so huge that when radial figures with equal and unequal spacing are compared, the question as to whether the equally spaced radial figure is set up vertically or not is of secondary importance.

Let us try altering by contraction and expansion a very simple image consisting of equal sizes and spaces. I am thinking of "orbit" B (50), consisting of equally broad, equally spaced parallel lines.

If we just gradually change the spacing, as we did in Questionnaire V with radial "orbit" A 2 with the equal spacings, we get a clear impression of movement and space. If we gradually change the equal spacings of parallels from close to wide, from left to right, (see Questionnaire VII) 96.4% of the test persons suddenly see movement, 65.7% of them in the direction of reading from contraction to expansion, while 30.7% see a movement from expansion to contraction, against the direction in which we read. If the direction of contraction to expansion runs against our reading direction, only 51.1% still see the movement as running from right to left, while 44.5% then follow the movement from expansion to contraction, in the direction in which we read. In both cases the majority see the movement as being from contraction to expansion. In both cases some 62% perceive the figure as "spatial", whereas in the case of the unaltered "orbit" of the equally space figure, 98.6% see no movement at all and 92% see the figure as "flat". Here, too, we perceive by comparison. The gradual alterations to the simpler figure lead to an impression of movement and space. We shall use other examples below to show that such simple perceptual conclusions play an important part in the decoding of our natural environment.

We have obtained from these experiments an initial insight into the thematic of even and uneven curvature, "contraction and expansion" and "enlarging and diminishing", i.e. of deviations from the schematism of simplicity, or, more precisely, of deviations from the orbit-like geometrical perceptual images – possibly developed some time after birth – which the infant draws from the end of its third year.

VII

Participants 137 — 7. Januar 2000

One tick above the line and one tick below the line for each figure.

65,7%	I see a movement from left to right	( )90
30,7%	I see a movement from right to left	( )42
3,6%	I see no movement at all	( ) 5
36,5%	I see the figure as more flat	( )50
62,8%	I see the figure as more spatial	( )86
0,7%	undecided	1

0,7%	I see a movement from left to right	( )1
0,7%	I see a movement from right to left	( )1
98,6%	I see no movement at all	(135
92%	I see the figure as more flat	(126
7,3%	I see the figure as more spatial	( )10
0,7%	undecided	1

44,5%	I see a movement from left to right	( )61
51,1%	I see a movement from right to left	( )70
4,4%	I see no movement at all	( ) 6
37,2%	I see the figure as more flat	( )51
61,4%	I see the figure as more spatial	( )84
1,4%	undecided	1

To avoid misunderstanding: I do not mean contraction and expansion and diminishing and enlarging as described by James J. Gibson (115), Boston, 1950, and as formulated by the gestalt psychologists after Koffka, 1951, especially by Arnheim in his book "Art and Visual Perception", 1954, in much more detail with the law of the simple image, according to which we see phenomena as spatial if the three-dimensional view makes them simpler, as is the case with all the texture examples described by Gibson. If we see 115 as spatial, the lines, variously spaced and of various sizes, will appear to have much the same size and spacing. This holds good for all the other examples Gibson produced in this connection. His conclusion was that altering textures gradually brought about a three-dimensional impression. Although this is right, he chose only examples which are better explained by Koffka's and Arnheim's more generally valid formulation. We are concerned at this point with examples of uneven curvatures, real contractions and expansions etc., which we recognize as expressions of movement and space, which do not, however, gain additional simplicity through being seen as three-dimensional.

We need discuss no further what Gibson described in 1950, since Koffka and Arnheim expressed it better at much the same time or slightly later. But in this connection Gibson said something which touches the problem I am addressing, and which I consider to be completely wrong. Since Gibson's book "The Perception of the Visual World"[26] is still haunting psychology courses and being read there, I am going to refer to the sentence and give my opinion: As far as the second [claim. The author] is concerned, it can be proved that the physical space between many things tends to be regular. This principle is true of grass in a meadow, trees in a wood, floorboards and the pattern on a carpet. It is amazing that Gibson names naturally growing things and artefacts in one breath; he is often wrong – in the case of the carpet pattern – and quite certainly in the case of his examples from nature.

Apparently Gibson is familiar only with well-tended putting greens. A normal meadow consists of all sorts of grass, growing to different heights, of broad-leaved "weeds", of flowers, of various tufts of vegetation variously spaced. They are certainly not regular and the spaces are irregular as well. A natural wood has both huge, ancient trees and saplings; there is constant variation in the size of the tree trunks and the spaces between them. Gibson seems to be thinking only of woods planted by forestry workers in much the same way as a farmer sows corn. But looked at more closely his claim does not even hold good for this semi-industrialized nature (111, 112, 256), and it certainly does not apply to natural life, to the original landscapes prevailing while our brain was evolving. Almost 100% of the answers to all our questionnaires result in seeing equal spacing and size as "stiff and not lively". Irregularity is a characteristic of life. It is no coincidence that perspective was discovered so late, not until the fifteenth and sixteenth centuries. It is not easy to find it in a landscape. It presupposes blocks of an equally large size and current straight lines, which rarely occur in nature. But let us return to our problem.

This time our starting point is dots assembled in a screen-like manner in Questionnaire VIII a contraction and expansion. We present two variations on this figure. In figure B the dots, organized in a screen, were displaced in such a way that they start with expanded spacing on the left, which contracts strongly on the right. The transition from closeness to wide spacing was managed so gradually that it is impossible to pick out clearly any group limitations. Figure C is a mirror-image of the whole process. This questionnaire was presented on 21 April 1995, before the students had been told anything about "contraction and expansion" or "enlarging and diminishing". Since the dots dealt only with contraction and expansion but not with enlarging and diminishing, in 1999 we put together another picture, consisting of equally spaced circles of one size (Questionnaire IX variously large circles, contraction and expansion). We then produced two variations on the picture, with large circles very close together on the left, gradually giving way to loosely spaced little circles but again in such a way that there were no borders around groups. We then produced a mirror-image of this picture. Once again we asked for a single answer per figure. The results of the two experiments can be compared. 90% of the 135 test persons saw no movement in the dots organized in a screen-like manner (Questionnaire VIII). When the circles were the same size (Questionnaire IX) this even increased to 92.7%. Most

[26] J. J. Gibson "The Perception of the Visual World". Boston, 1950

VIII

A

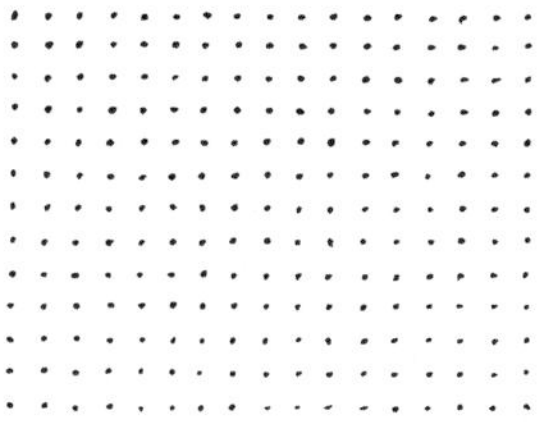

B

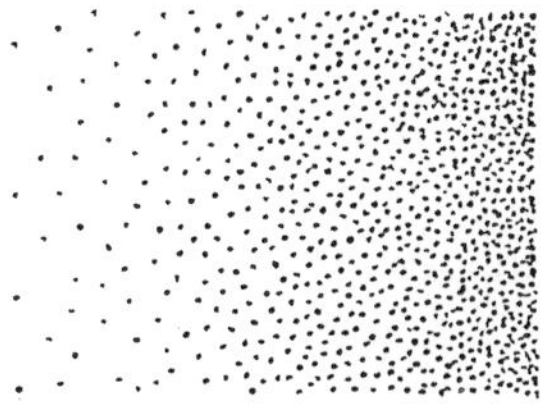

C

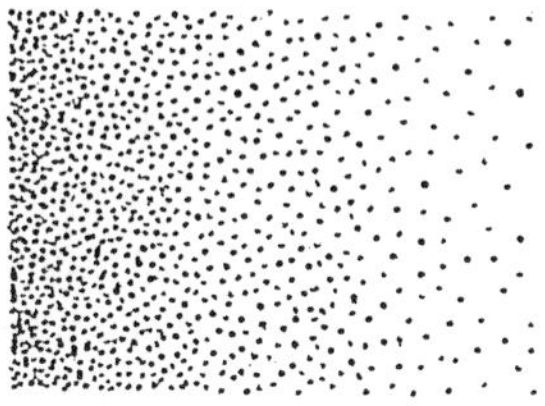

Evaluation on 25.4.95 in %

Choose one of the four possible answers for each of the 3 depictions A, B and C. Please tick just one answer for each picture. All four answers are available for each decision. – Please don't make up your own answers.

Number 135	A	B	C
I see a movement from right to left.	0 %	33%	49%
I see a movement from left to right.	0%	65%	48%
I see no movement at all.	90%	0%	0%
I cannot decide on a direction.	10%	2%	3%

XI

7.05.99

Please take a good look at the figures, then read all the possible answers and tick just one answer for each figure.

109 participants

Number			%
0	Do you see a movement from left to right?	( )	0%
101	Do you see a stiff and immovable pattern?	( )	92,7%
3	Do you see a movement from right to left?	( )	2,7%
5	Do you see an unclear movement?	( )	4,6%
109			

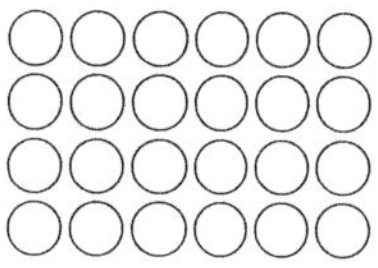

Number		%
70	Do you see a movement from left to right?	63,7%
1	Do you see a stiff and immovable pattern?	0,9%
35	Do you see a movement from right to left?	31,8%
3	Do you see an unclear movement?	2,7%
	[2 answers ticked in two cases]	
	One of these questionnaires could not be evaluated (right and left)	0,9%
109		

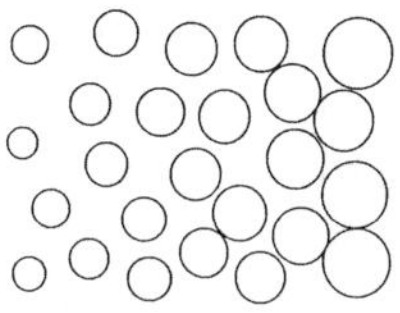

Number			%
39	Do you see a movement from left to right?	( )	34,8%
2	Do you see a stiff and immovable pattern?	( )	1,8%
64	Do you see a movement from right to left?	( )	57,1%
3	Do you see an unclear movement?	( )	2,7%
	[2 answers ticked in three cases]		3,6%

Two of these questionnaires could not be evaluated

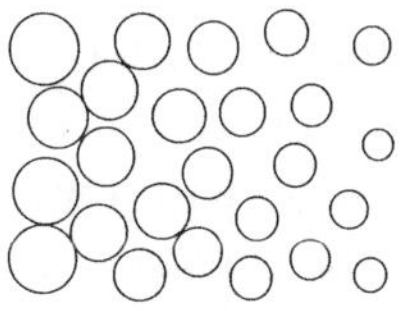

of the remaining 7% to 10% said "I cannot decide on a direction". There were just a few who thought they saw a movement from left to right in the circles organized in a screen-like manner.

It becomes increasingly clear that we believe the total correspondence between a figure in our environment with our visual cortex's schematism of simplicity to be stiff and immovable. We do not see vitality and movement until a conflict arises involving the simple images of our visual cortex, as in figures B and C. 98% of the test persons saw in figure B a movement from right to left or from left to right (65% from wide to close, in the direction in which we read, against this direction 49% from wide to close) (see questionnaire). In the case of the figure with variously spaced large and small circles, between 92.5% and 95% decided there was movement from left to right or from right to left. 63.7% saw the movement from wide to close and from small to large as being in our reading direction, while 57.1% saw the movement from wide to close and from small to large as being against it.

Two of the phenomena discussed previously are combined in the head of broccoli (104) to form one clear directional movement: a spiralling effect which starts off tightly curved at the top and then becomes gradually less curved as it reaches the bottom. This is true of the whole broccoli head as well as of each small individual pyramid forming the whole thing. The spiralling effect is combined with a gradual increase in size of the individual spheres which follow one another closely in the spiral. As a rule, we see the total direction of the head as moving from top to bottom beginning with small round shapes and ending with pyramids of increasing size. The small spheres up at the top are clearly shown in the photograph. But even the individual pyramids, increasing in size the closer to the bottom that they are, are made of small round shapes up at the top increasing to larger ones at the bottom. This effect is certainly strengthened here by the same structure being found in the whole head as well as in each individual pyramid-shape.

It would therefore appear that the alteration of closeness, size and curvature is experienced as the result of movement, even though, with these very simply constructed figures, the direction cannot be clearly determined. On the whole, rather more test persons chose the answer "from expansion to contraction" than the reverse. If we then add "from small to large" to "from expansion to contraction", the direction from wide to close and from small to large becomes rather more definite. But there are still so many persons preferring the opposite direction that we can only conclude that replacing equal with unequal spacing and size creates an impression of movement and space, but does not determine the direction seen.

Van Gogh's pen-and-ink drawing "The Village Street" (116) is a truly masterly recapitulation of our perceptions and the results of our questionnaires so far.

The changing density of the little parallel lines gives the thatched roofs on the left of the picture their plasticity and movement. In the right-hand bottom corner the vegetation growing upwards as the central point in uneven dynamic curves, with variously spaced radials, from close to loose, which seems to rotate around themselves, confirm our radial figures and the results on closeness and looseness and uneven curvature. In contrast to the extreme plasticity and dynamic of the roofs on the left and the vegetation on the right, we have the path to the sea and the walls of the houses, made of evenly spaced dots of the same size. Their screen-like appearance, lacking movement and perspective, only serve to increase the plasticity and dynamic of the roofs and vegetation.

As we go on, we shall continue to experience this. Everything I am addressing here, concerning phenomenological knowledge about sight was known to the great artists of the past. Art historians would not really have noticed it, had it not been for such instructive figures, constructed from geometrical forms, as Dürer's "Adam and Eve" or the man Leonardo drew in a circle and a square. We shall often return to this point below.

Since the test persons were to a certain extent influenced in their decisions by the usual left-to-right reading direction, both in the cases of the horizontally current alterations from expanding to contracting and from diminishing to enlarging, I set up another series based on dots organized in a screen-like manner (Questionnaire X centrally strewn dots). But this time we organized the process of expansion to contraction and vice versa from the centre to the edges. We asked the following questions about the three figures with dots of equal size: "Do you see movement from the centre

X

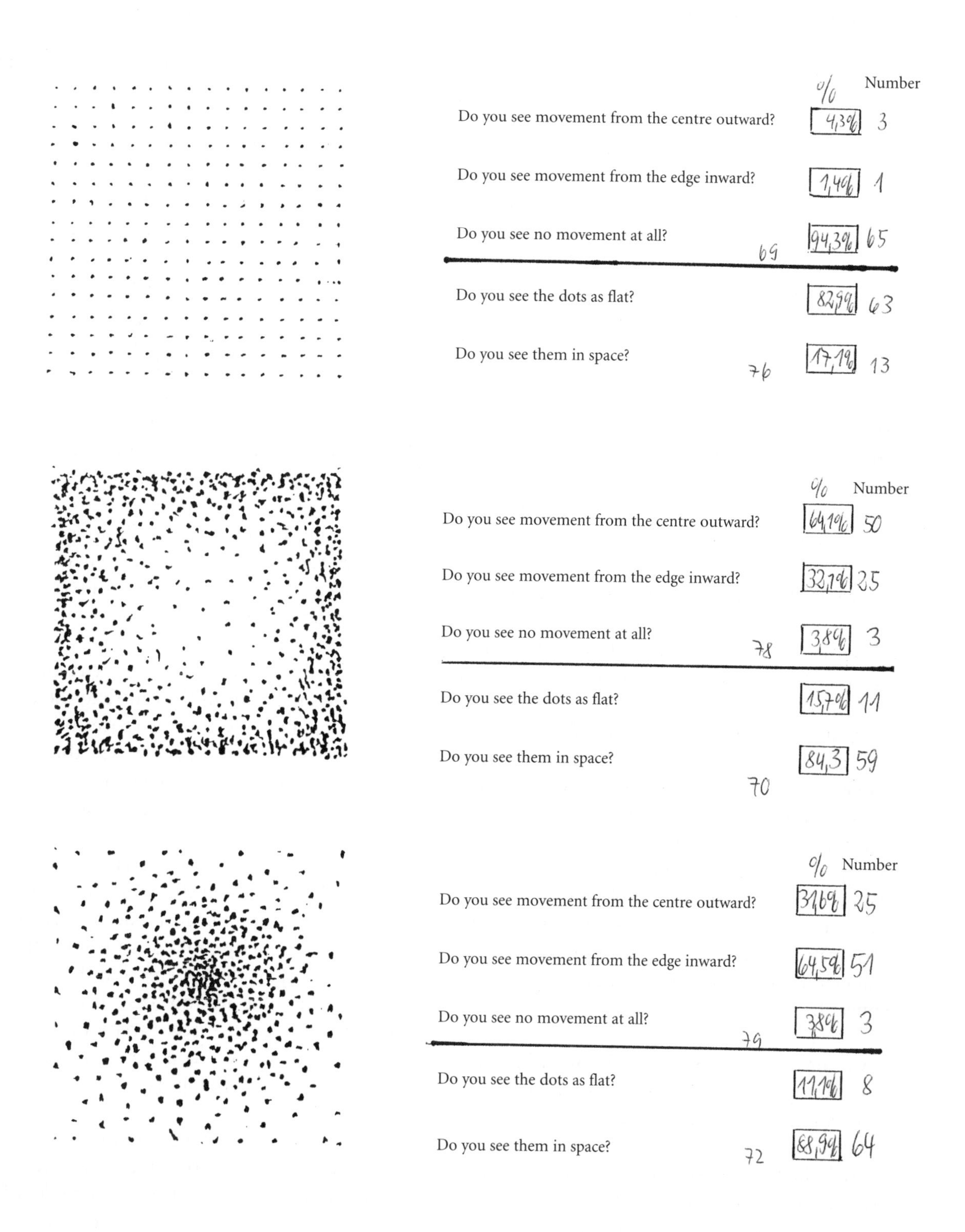
%
Number
Do you see movement from the centre outward?
4,3%
3
Do you see movement from the edge inward?
1,4%
1
Do you see no movement at all?
94,3%
65
69
Do you see the dots as flat?
82,9%
63
Do you see them in space?
17,1%
13
76
%
Number
Do you see movement from the centre outward?
64,1%
50
Do you see movement from the edge inward?
32,1%
25
Do you see no movement at all?
3,8%
3
78
Do you see the dots as flat?
15,7%
11
Do you see them in space?
84,3
59
70
%
Number
Do you see movement from the centre outward?
31,6%
25
Do you see movement from the edge inward?
64,5%
51
Do you see no movement at all?
3,8%
3
79
Do you see the dots as flat?
11,1%
8
Do you see them in space?
88,9%
64
72

5.05.1995

outward?" "Do you see movement from the edge inward?" "Do you see no movement at all?" This was followed in a new paragraph with another category of question: "Do you see the dots as flat?" or "Do you see them in space?" This time, in 1995, there were 78 participants, of whom 94.3% saw no movement in the screen picture. 82.9% saw the dots as flat, 17.1% as being in space. In the second picture, where the dots were widely spaced in the middle and closer at the four edges, 96.2% saw movement; 64.1% from the centre outward, from wide to close, and 32.1% from the edge to the centre, from close to widely spaced. 84.3% saw the picture as spatial, but 15.7% thought it was flat. 96.1% saw movement in the picture with the dots at their closest more or less in the middle. 64.5% from the widely spaced dots to the close ones, from the edge inward, but 31% saw a movement from the centre outward. In this case about two thirds of the test persons saw the direction of the movement from wide to close spacing and a third saw it the other way round. 88.9% of them saw the figure as spatial and 11.1% as flat.

Yet again a deviation from the "schematism of simplicity" – in this case a deviation from a screen-like even distribution of dots of the same size across a surface – is seen as resulting from movement. Let us recall that a six-week-old baby recognizes a chessboard pattern and a three-year-old child draws a screen or ladder system. Obviously a deviation from the screen is connected with a spatial effect, although, for example, the contraction in the centre is not made more simple by spatial understanding, so that the gestalt psychologists' law of the simple form would not explain this result. We shall have to return to this observation, because it will lead to an important extension of gestalt psychology's signification concept.

If we replace the dots of one size with increasing and diminishing circles (Questionnaire XI centrally strewn circles) in the centripetal and centrifugal contraction and expansion we have just discussed, the test held on 19 November 1999 showed that once again the majority (the percentages were much the same) saw movement from wide spacing to closeness. It made no difference if the large circles were widely spaced and the small circles close together or vice versa.

In this case the direction of the movement seems to depend primarily on contraction and expansion and not so much on varying sizes. These can perhaps be understood as a result of distance – and thus in agreement with the law of the simple form – even though they are not consistently constructed perspectively. However, this could change if the figures in the questionnaire clearly excluded a spatial reason for the alterations in size. We already have a first impression of this in the case of the alterations unifying themselves with the horizontal displacement from loose to close, going from small to large circles. However, this idea must be confirmed by more definite proof.

But first we are going to concentrate on circles or spheres, whose varying sizes depend essentially on varying distances. We fill a large circle (Questionnaire XII flatly and spherically organized circles) with many little circles of one size, arranged in concentric rings, that is as evenly as possible. Since this is not quite possible mathematically/geometrically we have to resort to a "trick". In the centre the spaces between the circles will have to be a little larger; but hardly anyone will notice it. On 23 April 1999 some 77% of 137 students tested saw the image as "flat". 17.5% saw it as alternating between flat and three-dimensional.

In figure B these spheres, drawn as circles, became uniformly smaller from the centre to the circumference; so, gradually, did the spaces between the spheres. 95% of the test persons saw this figure as "three-dimensional". Presumably they saw a large sphere made of evenly spaced smaller uniform spheres. This interpretation would seem to be based on gestalt psychology's law of the simple form. In this way – that the whole figure was seen as a sphere and not, like its predecessor, as a circle – so much simplicity could be gained by means of the uniformity of the spheres and the spaces between them, that in this case our perception obviously prefers to see three-dimensionality rather than flatness – completely in accordance with the gestalt psychologists.

We then produced a third variation on the same figure. We shifted uniform spheres, which were on the whole sphere's circumference or circles of latitude, in such a way that we achieved a symmetrical pattern consisting of three centrally meeting circles of longitude or six uniform radials radiating from the centre. Between them lay triangular circle compositions. What we had, therefore, was a radial figure with evenly spaced radials, diminishing outward uniformly six times. But the progres-

XI

Participants 95

19.11.1999

One tick above the line and one tick below the line for each figure.

I see movement from the centre outward	(6)	6,3%
I see movement from the edge inward	(1)	1,1%
I see no movement at all	(86)	90,5%
No ticks	2	2,1%
I see the circles as being arranged in a flat way	(84)	88,4%
I see the spheres as being arranged in space	(11)	11,6%
No ticks	-	

I see movement from the centre outward	(29)	30,5%
I see movement from the edge inward	(55)	57,9%
I see no movement at all	(10)	10,5%
No ticks	1	1,1%
I see the circles as being arranged in a flat way	(19)	20,%
I see the spheres as being arranged in space	(75)	78,9%
No ticks	1	1,1%

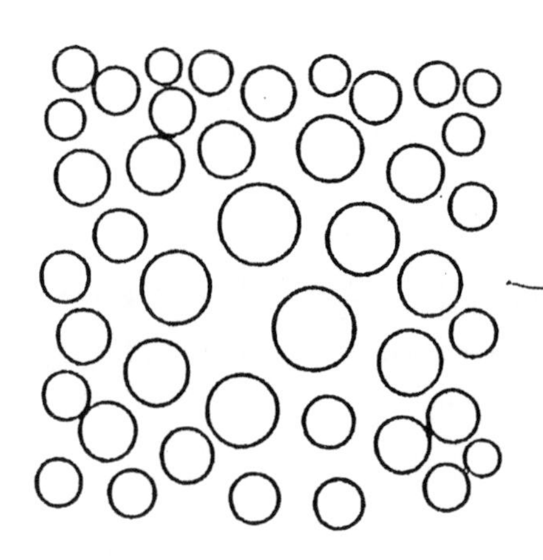

I see movement from the centre outward	(61)	64,2%
I see movement from the edge inward	(28)	29,5%
I see no movement at all	(5)	5,2%
No ticks	1	1,1%
I see the circles as being arranged in a flat way	(13)	[illegible] 7%
I see the spheres as being arranged in space	(81)	85,2%
No ticks	1	1,1%

XII

12.11.99

Please take a good look at both the figures, possibly holding them at arm's length, then read all the possible answers and tick just one answer for each figure.

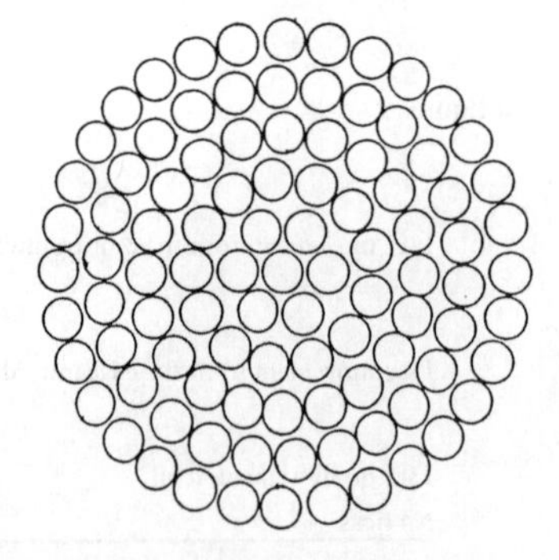

Do you see the figure as flat? (111) 77,1%

Do you see the figure as three-dimensional? (20) 13,9%

I cannot decide. (13) 9%

144

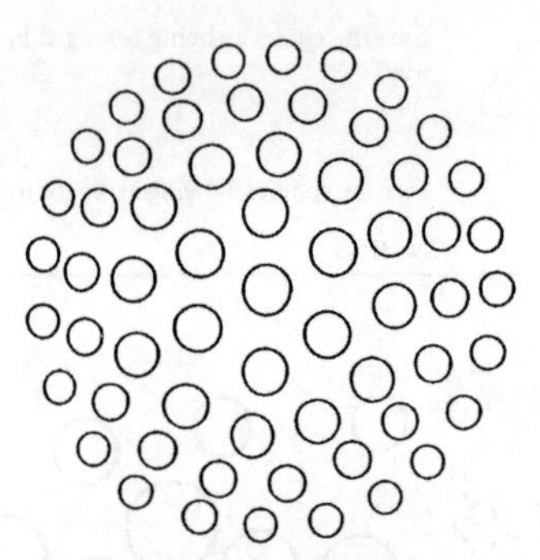

Do you see the figure as flat? (2) 1,3%

Do you see the figure as three-dimensional? (141) 98%

I cannot decide. (1) 0,7%

144

Please take a good look at the figure, possibly holding it at arm's length, then read all the possible answers and tick just one.

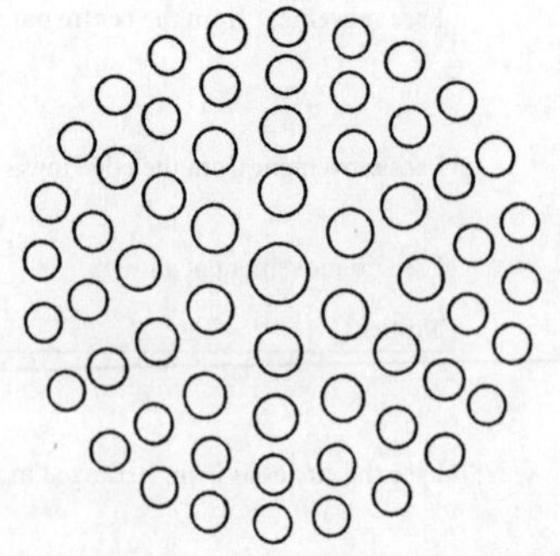

Do you see the figure as flat? (71) 48%

Do you see the figure as three-dimensional? (51) 34,4%

I cannot decide. (26) 17,6%

148

sion of large spheres in the centre to smaller and smaller ones at the periphery and the corresponding alteration of the spacing stayed the same within this pattern made of six radials or three meridians as it was in the previous figure B. So it was entirely possible, in accordance with the law of the simple form, to see this image, like figure B, as three dimensional. It would then be a sphere observed from its North Pole, so that the longitude running to the equator would have to form a symmetrical figure. In this case, too, the separate spheres making up the whole sphere, and the spaces between them, would get symmetrically smaller. Despite all this, 76.6% of the test persons see this figure "alternating between flat and three-dimensional", some 14% see it only as "flat" and 9.5% as "three-dimensional". So the decisions for flat and three-dimensional are almost equally balanced. So this image "rocks", even though the symmetrical pattern made by the three meridians would definitely be simpler if seen as three-dimensional rather than flat, so far as the size of the individual spheres and the spacing between them are concerned. Why, then, does the figure tend to be seen more as flat?

I have already put forward my idea that we have developed, as a perceptual image, a symmetrical ray or radial figure, such as those drawn by children in their third year, shortly before they invent the tadpole. Hoffman and others demand it as an "orbit" from birth onward. In my opinion, all perceptual images are two-dimensional. Although our third figure consists of spheres which change from large to small, as do the spaces between them, thus conforming to the laws of perspective, the two-dimensional radial structure is so dominating that, despite the gain in simplicity the question as to whether the figure is three-dimensional is answered by saying that the image as a whole is balanced between the flat and the three-dimensional. Into this category we can place the sign of the German Catholic Bishops' Conference, made of cruciformly organized, gradually diminishing circles.

This similarity to the radial "orbit" would also explain the two-dimensional effect of the hexagonal, symmetrical cube, drawn as transparent (60 c). This could also be the reason for the figure "rocking" between this hexagonal pattern and the three-dimensional cube. Whereas, for instance, the spatial effect of the uniformly large dots gathering towards the centre could not be explained by the law of the simple form (Questionnaire X).

If we proceed from the idea that our perceptual images, with which we compare our visible environment, are, firstly, two-dimensional and, secondly, always as simple as possible, – i.e. comprising uniform lengths, angles, curvature, density and size – we can go on to say that, in accordance with the law of the gestalt psychologists' simple form, all deviations from these simple structures will be seen as three-dimensional, if this brings about an increase in simplicity. But if simplicity is not, or only partly increased, we see these deviations in many cases as the result of movement, and so as three-dimensional, because movement always happens in space. Because the uniformly large dots get closer together toward the centre, we perceive a movement in that direction and thus, also, a three-dimensional gathering together. But an increase in density at the edges of the square also creates a spatial impression, at least a view in the centre, especially compared with the screen-like organization of the dots.

But although the unevenly bent curve of a drop or vessel-form deviates from our visual cortex's circle or rectangle structures, perception as three-dimensional makes it not simpler but more complicated (Questionnaire XIII). Despite this, 54.4% of the test persons see such a curve as spatial rather than flat, while only 34% see it as flat. Only 16.7% see a circle as spatial, that is, as a sphere, while 77.4% see it as flat. We tend to interpret the difference between the unevenly bent curve and the flat circle as a directed expansion, which makes us see the image as both dynamic and spatial. That is why dynamic forms, even in their spatial effect, often completely dominate geometrical ones. Arnheim also got similar results questioning his students after I raised the subject in the 1970s.

Often, in our perceptual reality, three-dimensional understanding, resulting from the related gain in simplicity, is unified with the three-dimensionality resulting from the movement deforming the simple form, and which we can discern in it. Particularly simple examples of this are Doric columns or cylinders unevenly curved at top or bottom. The ellipses seen from all side views of the circular cross-section of the column or the cylinder lead us, because of the simplifying effect, to a three-dimensional understanding, which is increased to vital three-dimensionality by the

XIII

10.11.1999

Please look at the figure closely, carefully read through the possible answers and just tick one.

Participants 103

I see the figure as a sphere	(56)	54,4 %
I see the figure as more flat	(35)	34,0 %
I cannot decide	(12)	11,6 %
Participants	103	

I see the figure as a sphere	(17)	16,7 %
I see the figure as more flat	(79)	77,4 %
I cannot decide	(6)	5,9 %
Participants	102	

expansive effect of the unevenly curved vaulting (61 a, b, 109).

The human figure is an especially complicated example of a three-dimensional effect arising from the combination of the law of the simple form and the deformation of simple forms experienced as movement and three-dimensionality. In this case, however, the three-dimensional effect based on dynamic deformation is clearly dominant. The three-dimensionality arising from the figure's composition, which compared to the axial-symmetrical standing figure with hanging arms (42), leads by means of reaching movements, bending limbs and the attendant shortenings, i.e. to apparently various lengths between thighs, calves or arms – this three-dimensionality is based on the law of the simple form. Understood as three-dimensional, these variously long limbs become the same length again, i.e. symmetrical and thereby simpler. But the whole spatial effect of the corporeal details of the legs, arms, trunk and head is mainly based on a dynamic deformation of the basic forms, which we see as a process and thereby as three-dimensionality (117). But perception of the human figure depends on such a complicated combination of various categories that it cannot be exhaustively dealt with at this point; it will appear in a later chapter.

Chapter 8
The Classification of Memory Pictures by Students. Reproduction Memory – Identification Memory

At this point I think we should look at the subject from the opposite point of view. We should no longer ask the test persons to choose between drawings and answers prepared by us. Instead, we should set them a subject and ask them to make their own drawings. I know that there is a great deal of reservation about employing this method, because of claims that, when it comes to drawing, people are so very differently gifted and, particularly, that their education in the subject has been to such different standards, not to mention the different conventions that exist. My answer to this is talented, above-average draughtsmen are so rare that they can be ignored in a statistical examination, and that far more than 90% of present-day German school-leavers have received no education in drawing at all, the reason being that most art teachers are unable to draw. The general knowledge of German school-leavers is of a similarly low level, as we have shown using other questionnaires[27].

If we set the drawing test in such a way that it needs no drawing ability and conventions to solve it, if we just aim to find out whether or not people retain images of objects in their environment, then the proposed test has far clearer results than most other methods. This is true, even though in 1984 Irvin Rock wrote that there are "probably more suitable methods" to "test how a figure is retained in the memory". "More suitable methods" are certainly not comparisons of size in the mind, such as Paivio (1975) demanded from his test persons ("Which is bigger, a cat or a stag?") Nor are they students' "mental rotations", the duration of which Shepard and his assistants have been timing with a stopwatch since the seventies. The claim made by Finke (1980) that "on an initial level" ... "perception and imagination lead to the same results" shows how little suited such tests are to elucidate the relation between perception and imagination. The culmination of this is the claim that "the conscious imagining of an elephant" leads "to the same results as the perception of an elephant". No, that really is not the case, as our drawing tests will prove beyond doubt. And over and above this, they will interestingly confirm our tests and ideas to date. So let us simply put it to the test. The result will show how groundless are these doubts concerning drawing tests, if they are properly conducted.

We used the first lecture for the new intake of architecture students in 1998 to do this test: after distributing pencils and paper to all the students, we asked them to imagine and draw a spruce or fir-tree, also imagining a section through the tree from top to bottom. I also said that the drawing should show how, and in which direction, the branches grew out of the vertical trunk. I asked them to draw only one tree, not to put in any shading but just to use lines. There was plenty of room to work, as 166 students participated in the first test and 169 in the second (three arrived too

[27] Jürgen Weber "Abitur gleich Hochschulreife?", in "BDK Mitteilungen", Fachzeitschrift des Bundesdeutscher Kunsterzieher e.V. 1/95 p. 25-27

late) in a lecture hall intended for 300 to 400. We did not set a time for completing the test; we did not collect the papers until everyone was ready. For the first test this took about 15 minutes.

Following this, I asked them to draw a palm-tree, again with a vertical section through it. The drawing should again show how and in which direction the leaves grow out of the trunk. Again we asked the students not to use graphic shading or effects and to draw only one tree. This time the students were a little quicker, about ten minutes.

Let us start with the fir-tree: 161 students, almost 97%, drew all the branches from top to bottom parallel to each other, evenly spaced, growing out of the trunk at the same angle. The biggest group among these (74 students, 47.1%) had the branches not only evenly spaced but growing out of the trunk at right angles – i.e. horizontally (126, 127). 25 of them did not draw any needles. During a probationary lecture a sculptress, some 40 years old, introduced a fir-tree modelled on exactly such simple lines and cast in bronze (125) as one of her major works. She was applying for the chair of full professor (C4 in the German educational system) at the Technical University of Braunschweig as my successor. This did not upset anyone. When I pointed out how primitive it was, none of my architecture professor colleagues understood why there should be any objection. The appointment board placed her first on the shortlist of the three most appropriate candidates. Due to certain circumstances, this list was, however, withdrawn.

The second biggest group (66 students) drew all the evenly spaced, parallel branches angling downward from the trunk (123, 124, 128, 130), while 20 (the third largest group) had all the branches pointing upward at the same angle, parallel and evenly spaced (129). They drew the needles at the same acute angle, pointing towards the end of the branch. So almost 97% of these students had no memory of the fact that the little branches at the top of spruces and firs grow steeply upward and that as the branches grow larger and heavier they are first horizontal and then bend toward the ground, as the photograph from nature shows (132). This is a basic principle of growth and can be observed in most large trees and bushes. It is particularly clear in firs, spruces and most palm trees.

The alteration of the angle is usually accompanied by increasing and decreasing density. At the top of a tree whose branches get longer from top to bottom, but are all very light in relation to their stiffness, they grow fairly parallel upward and are rather dense. In the second third they get larger and heavier, sink down to the horizontal and finish up pointing downward. In this section of the tree the branches are much further apart and follow a rotation schema of changing angles and densities, such as we have already examined in our point symmetrical radial figures. In the lowest third there is little increase in the length and weight of the branches – they have reached their full size – so that this downward alteration of their angle of growth slows down and, through the increasing volume of the twigs, a considerable density recurs. Apart from this, none of the 161 students noticed that branches usually grow from the trunk with uneven curvature. Some drew straight branches, others evenly curved segments of circles (cf. our examination of unevenly curved lines). But, as I have already said, this is the way it is nowadays with professors of art and architecture, too – nor are things any better when it comes to students and professors of garden architecture (questions put to garden architects from Essen in 1993).

Almost 47% of the students used the simplest possible schema; only straight lines or segments of circles, rectangular joins and even spacing. They fitted the fir-tree image into the ladder or screen-schema recognized by six-week-old babies and scribbled by infants in their third year. The image that results is so universal that it could also represent many other objects. Nor must we forget that we are dealing here with students starting to study architecture, who had to average 2- (B- equivalent in the British system) in their school leaving examinations: more than half (52%) drew the fir-tree's branches angling evenly upward or downward, evenly spaced and straight or evenly curved. This is the way children draw a fir-tree from their seventh year at the latest. So altogether almost 97% of the students had stored firs and spruces in their optical long-term memory in the two simplest images possible – though there was one student who went one step further and drew his fir-tree as a simple triangle on a stick. There was only one woman student (0.6%) who drew her fir-tree with full rotation of the angles, but with no changes in the spacing and no uneven curvature of the branches (131). Since she did not later prove to be particularly gifted, we are left with the question of how she arrived at this result. In any case, she and the slightly more than 2% whose drawings showed signs of rotation do not alter the result that normally objects in our environment are stored in the simplest geometrical way in our long-term memory. (In a later chapter we shall report on my students drawing animals and faces from memory.) I would also like to comment on the illustrations and say that the overall picture presented here is much too positive. This is because the one rather good fir-tree is portrayed along with only a few examples of the many drawings lacking in information.

It is possible to raise the objection that, by asking the students to draw not a three-dimensional fir-tree but a vertical section through one, we obliged them to produce such schematic drawings. But the direction of the branches and the spacing between them are the most important preconditions for a realistic, three-dimensional drawing, too, besides which, this two-dimensional request fits in well with our optical memory, makes things easier for our visualization.

On 10 February 1993 we showed 316 engineering students the slide of a fir-tree for 30 seconds and then, after we had divided them into two roughly equal groups, gave each group one of two variations on a questionnaire. One group had to answer three questions, the other seven. The first question for both groups was, "Which geometrical form is the closest to this fir-tree?" 57.5% suggested a triangle, a two-dimensional form, even though they had just seen the photograph of the branches growing round the trunk and had plenty of time (15 minutes) to think about the question. On top of this, these were engineers, who studied geometry. 45.2% answered "cone". I believe that if we had carried out the test as a quick question-and-answer game, the percentage answering "triangle" would have been even greater, but it is very difficult to organize such a game under proper conditions with a group big enough to be of use statistically. But does not almost everyone find the answer "cone" rather unnatural? I myself think first of a triangle when asked to name a simple geometrical form resembling a fir-tree. In any case, 57.5% "triangle" answers from students who also have to pass geometry examinations, and who were able to reflect for some time on their answer, indicates that I seem to be right in thinking that our form schemata are two-dimensional. This is also indicated by the two-dimensional children's drawings of fir-trees, which we obtained from 25 schoolgirls on 22 July 1999 (118 – 122). Of course, unlike the architecture students, we did not ask them to produce "section drawings", but just asked them to draw a spruce or fir-tree (cf. seven-year-old Constantin's drawing), (43). We probably met halfway our students' ability to imagine objects by asking them to produce a sectional drawing. The percentage of simplest and simple schemata representing fir-trees in the students' drawings probably came very near, at least basically, to what our visual memories can produce.

In this connection it is certainly interesting to find out how the test persons themselves judge such drawings:

On 26 November 1999 we set a questionnaire in front of 119 students. It showed three schema drawings of a fir-tree (Questionnaire XIV), A, branches joined on only at right angles, B, branches joined on at an acute angle, pointing upward, as were the needles and, C, all the joins rotating, from the steeply upward angled little twigs at the top, through the horizontal ones, and finishing with the downward-pointing large branches, also passing from dense, to widely spaced and back to dense. We added rotating lines to the branches, their angles varied and they got shorter toward the tips of the branches. These represented, in the form of a rotation symbol, not only the twigs, but also the needles on twigs and branches. We depicted the branches, growing out of the trunk in rotating angles and changing spaces, with their main curvature starting directly at the trunk. In reality the main curvature can be at either the beginning or the end of a branch. The latter shows how right we were to say that, under certain circumstances, unequally curved lines can also be read from the main to the flat curvature. Besides this, categories frequently encompassing two possibilities are needed to comprehend the variety of natural phenomena. The categories are mutually dependent and only then become clear.

Not even the last schema described is drawn as either naturalistic or three-dimensional, as comparison with the photograph from nature will confirm. We omitted any depiction of the way branches, depending on wet and dry years, grow

XIV

Participants 119

26.11.1999

For each depiction of a fir-tree, please tick the description which you find to be the most fitting when comparing the drawings. Please choose one description above the line and one below.

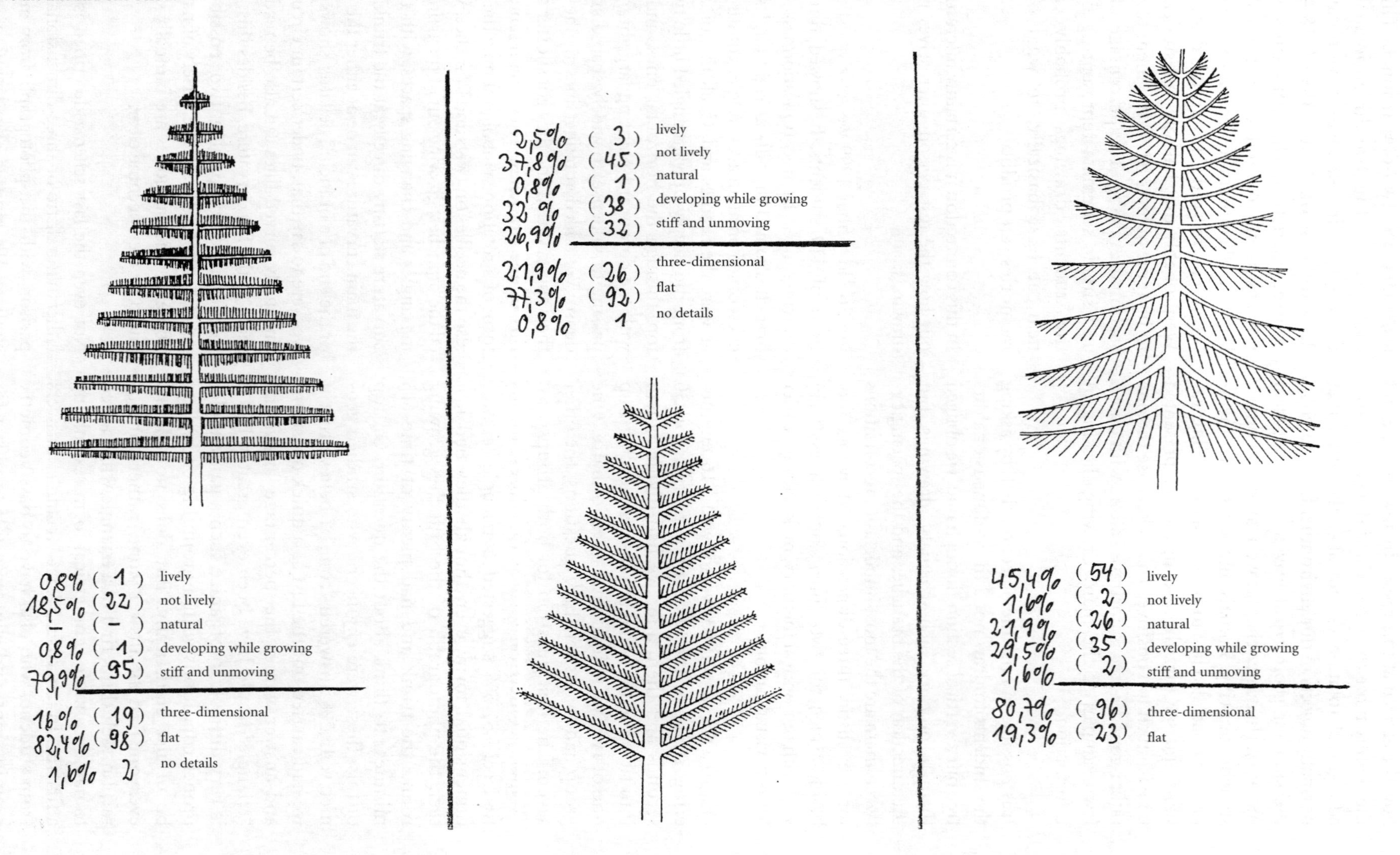

to different lengths, especially in the upper reaches. This makes the result of the questioning even more striking.

For each of the three drawings the students could choose one of five descriptive words for the tree, and under a horizontal line they could choose either the word "three-dimensional" or "flat" (Questionnaire XIV). Three of the descriptive words had very similar meanings: "lively", "natural", "developing while growing". Two, with opposite meanings, were also very similar: "not lively" and "stiff and unmoving".

98.4% judged drawing A, of a fir-tree with horizontally growing branches only, to be "not lively" and "stiff and unmoving". 82.4% said "flat" and 16% "three-dimensional", although 47% of the same students had drawn such a fir-tree.

In the case of drawing B – a fir-tree whose parallel branches angled upward – 64.7% said "not lively" and "stiff and unmoving" and 35.3% said "lively", "natural", "developing while growing". (The extreme variations in the division of the percentages among the last three choices (see questionnaire) shows how right we were to choose different words with a very similar meaning.) Here 77.3% said "flat" and 21.9% "three-dimensional". As a total result the sloping branches were accepted as more lively and moving than the ones at right angles and also rather more readily seen as three-dimensional than in the case of drawing A.

In the case of drawing C – a fir-tree whose rotating branches are slightly unevenly curved – 96.8% chose "lively", "natural" and "developing while growing". 80.7% thought the drawing was "three-dimensional" and 19.3%, "flat".

It is worth considering why, in the case of C, only 19% chose "flat", but 80.7% chose "three-dimensional", even though we used none of the usual methods to indicate it in our schema. There is no perspective, no cross-sections indicating that it is three-dimensional and, above all, there are no intersections indicating this, nor is it easier to understand if three-dimensional.

Obviously, once again our visual cortex interprets big deviations from the "schematism of simplicity" as three-dimensionality, or, to express it in another way, the expression of movement to be seen in the frequent choice of "lively", "natural" and "developing while growing" seems also to be understood as three-dimensionality. We have already pointed this out in our earlier experiments. For our perception, movement can take place only in space. As soon as even a minimum of intersections is added, almost 100% choose "three-dimensional".

On the same day we set a questionnaire with three schematic drawings of pinnate leaves, using much the same system as we had with the firs (Questionnaire XV). The same expressions were to be used to judge these drawings. In the cases of A and B the results were so similar that they need not be examined more closely (see questionnaire). Only in the case of C did considerably more students choose "three-dimensional" than for the fir-tree C: 99.2%. And indeed this was not a schematic drawing, but a tracing of a photograph from nature (134). But in the case of this really very flat young fern-leaf there were very clear intersections where the individual leaf segments sprang from the central nerve. Even more than perspective, they are the most important indications of three-dimensionality. Apart from this, we find here all the other criteria we have mentioned so far: rotating angles ranging from steeply upward to almost horizontal, dense to wide spacing and unequal curvature, in this case outward. Here it is read from the flat to the strong curvature. So this picture combines the mainly preferred reading direction from flat to strong curvature with all the other movement criteria to form a fan-like developing growth process. This leaf obtained the highest possible choices of "lively", "natural" and "developing while growing": 100%. This result was helped by the natural asymmetry of the photograph, which also depicts a deviation from the schematism of simplicity. Our three schematic drawings of firs are all strictly symmetrical.

At this point we should mention that in the test held on 23 October 1998, the drawings of palm-trees from memory were at least as poor as the fir-trees. Almost 50%, to be exact 49.07%, drew the palm with opposing, evenly spaced leaves, like leaves growing on a twig (135,136,137). There were also two students who drew sets of two opposing palm-leaves, like a pinnate leaf with a central nerve (144). That was a total of 51.2% who did not even remember that as a whole palm-leaves look like a spherical body, as the photograph shows (133). But the ancient Babylonians, at the time of Assurbanipals (669 – 627 BC), frequently depicted

XV

Participants 120 26.11.1999

For each of the three depictions of pinnate leaves, please tick the description which you find to be the most fitting when comparing the drawings. Please choose one description above the line and one below.

(11)	lively	9,2 %
(41)	not lively	34,2 %
(10)	natural	8,3 %
(37)	developing while growing	30,8 %
(21)	stiff and unmoving	17,5 %
(20)	three-dimensional	16,7 %
(99)	flat	82,5 %
(1)	without details	0,8 %

	( )	lively
20,8 %	(25)	not lively
	( )	natural
	( )	developing while growing
79,2 %	(95)	stiff and unmoving
4,2 %	(5)	three-dimensional
95 %	(114)	flat
0,8 %	(1)	without details

(38)	lively	31,7 %
( )	not lively	
(21)	natural	17,5 %
(61)	developing while growing	50,8 %
( )	stiff and unmoving	
(119)	three-dimensional	99,2 %
(1)	flat	0,8 %

palm-trees in this way, conforming to the ladder schema, with opposing leaves (145).

About 39% drew the palm-leaves as a circle, but evenly spaced like a windmill, or occasionally with fortuitously different spacing, which, however, gave no indication of a growth principle as described for the firs (139, 140). Depictions of palm-trees following such radial orbits can also be found at the time of Assurbanipals (146). Apart from this, these depictions of palms also indicate that we probably have a rosette image for purposes of geometrical comparison. After all, palm-leaves have pointed and not rounded ends. So we are looking at the simplest geometrical schemata, drawn unaltered from the lively impression left by the ubiquitous palm-trees.

3% drew the palm as if it were a weeping willow, with branches hanging downward (143). Only one person drew the palm still with opposing leaves but with so many of them that the image took on a look of an ellipse. The leaves were clearly bent, as they are in the lower reaches of a palm-tree (142). One (some 0.7%) drew the palm with rather circular radials, somewhat more widely spaced at the top and thus with certain rotational elements (141). Roughly 2% drew their palms with only three evenly spaced leaves (144).

The conclusion is that more than half the students had not even grasped that palm-leaves are usually organized spherically, more radially, so if depicted in a section the total impression is circular. They did not remember that, like firs, the central palm-leaves are mostly more widely spaced and denser at top and bottom. Nor had they seen that the top palm-leaves are relatively straight and sword-like, while the lower ones curve more and more strongly downward, finally bending toward the trunk – they depict a growth process. Half of the students used the most primitive schema of leaves growing opposite each other. There were 169 participants.

An important result of the test drawings is – and we shall be quoting further examples – that our visual long-term memory geometrically simplifies the idea of objects in our environment to fit in with existing simple schematism. As I have already said, I shall call this memory as a part of the "declaration memory" the "reproduction memory". The next experiment will show how hard this reproduction memory clings to simple geometrical forms, or how quickly, contrary to the opinion still held by many perceptual psychologists, it forgets nature's more complex structures.

One week later, on 30 October 1998 – at that point the students had been told nothing about the result of their drawings of firs and palms – I told them at the beginning of the lecture that I again wanted them to draw these trees from memory, again as a roughly vertical section, and also a young fern-leaf, but not until I had projected these three plants on the wall for half a minute. So for exactly 30 seconds we showed the fir-tree (132), distributed paper and pencils and asked for silence. After about a quarter of an hour we collected the drawings and darkened the room again, so we could project the picture of a palm-tree (133) onto the wall for 30 seconds. The students then started to draw again and finished after about a quarter of an hour. We then projected the picture of the fern-leaf (134).

Not only had we just shown the students slides of the objects we wanted them to draw, but only one week earlier they had drawn palms and firs from memory, so had already addressed this question. We should have been able to assume that they would approach the projected images with particular questions in mind. The result was an improvement on the first test, but still very schematic. There were 157 participants. In this test the students were able to take advantage of the short term-memory.

This time 15 participants drew circular treetops on their palms, with clear rotational elements, such as different spacing from top to bottom or varying curvature, ranging from almost straight at the top to strongly curved or bent at the bottom. That was 9.6%. The best by a long way was (147), followed by (148) and (158). The biggest difference in the depictions after the students had looked at the slides was the decrease in the amount of paired opposing leaves. Only 14 participants still produced these, not quite 9% (149, 155, 160) and 16 students, 11%, again drew the palm like a weeping willow (150). But 71.3% depicted the palm as a more or less spherical or circular schema, though with roughly even spacing, like a windmill, and gradations of curvature. So a large majority had abandoned the opposing schema in favour of a radial schema, though retaining the even spacing and gradations of curvature. So looking at the slide had had an effect (151 – 154, 156, 157), but the short-term memory was obviously not able to retain complicated connections. 91% depicted the palm schematically either with radial or opposing leaves, or drooping evenly downward like a weeping willow. Only 9.6% – but that was an improvement – had recognized rotational elements and so produced drawings close to nature.

Similarly, the drawings of fir-trees also produced better, but still very schematic results (157 participants). The group depicting branches evenly spaced and angled, as straight parallel lines or segments of circles, was reduced to 67.11% (165 – 168, 171, 172). But 25.6% – apart from the rotating angles, as in the case of the point symmetrical radial figure B (Questionnaire V) – had drawn the fir-tree without different spacing (170 was by far the best). That totalled almost 94%. But, as I said, 25.66% had at least realized the transition from straight branches growing upward at the treetop to those bending towards the ground at the bottom. This time 7.23%, 11 students, drew the fir with rotating angles and uneven curvature, but retained the even spacing. One student drew the fir without rotating elements and with even spacing but uneven curvature (169).

Rather worse, very strongly simplified schematic drawings resulted from projecting the fern-leaf (134, photograph from nature) for 30 seconds. This was probably because the students were drawing this motif for the first time.

Through this geometrical simplification, 20% of the drawings of the fir-tree and the fern-leaf (161, 162 fir and 163, 164 fern-leaf) could be mistaken for each other. In both cases the fir-tree and the fern-leaf had been fitted to the ladder or screen-system, which led to such generalized images that, in the drawings of many students, it was no longer possible to distinguish the two very different plants, fir-tree and fern-leaf. Let me reiterate that we were not dealing here with a particularly untalented cross-section, but with architecture students, who later, after graduating, would be designing buildings for completely different purposes in accordance with one and the same geometrical schema they had used for fir-tree and fern-leaf.

Of course, it is possible to object that architecture and fine art used to offer far greater variety of form. Of course, in those days the reproduction memory, and with it the ability to invent forms, was systematically schooled and developed very early on. This certainly led to the self organization of the brain developing in a different way (Singer). But if that is not done, as is the case today, the reproduction memory does not progress beyond the stage reached by a ten or twelve-year-old child, see the drawings of the elephants and runners as well as the portraits of fathers (282 – 297, 416 – 431, 432 – 445). And that, in turn, affects our ability to perceive differentially. One of the most important human mental abilities is degenerating apace.

Another proof of this is the fir-trees I asked some 30 schoolchildren between 7 and 11 to draw from memory. The drawings of most of my architecture students correspond roughly with this level. The primitive geometrical schematism of the depiction is very similar. The unpractised reproduction memory probably corresponds to equally simple seeing. Visual memory and seeing need practice. Contemporary educational politicians completely ignore the fact that this is an important part of creative thought. They believe in the computer just as our forebears believed in the Good Lord.

It is quite clear that in our reproduction memory – also in our short-term memory – the complicated objects in our environment are very quickly fitted into simple geometrical structures. This is not contradicted by the fact – in any case not statistically – that fine artists can reproduce from their imagination more complex images than can others. But even their pictures are variously complex when compared to natural forms and – even if rather differently – almost always geometrized. We shall return to this in the second part of this discussion.

In commenting on our questionnaire with the three variously complex schema drawings of a fir-tree I pointed out that we had not asked for a depiction of the different lengths of branch resulting from wet and dry years. This area was addressed by another test. On 11 November 1994 we set 134 students a questionnaire with three figures consisting of six parallel straight lines each (Questionnaire XVI) to find out in which direction they thought the lines were pointing. In the case of the six equally long lines assembled in a block (one of Hoffman's orbits), 92.5% said "in no direction". But when it came to the lines of various length, organized in the manner of an arrow or triangle,

XVI

In which direction are the 6 lines pointing?

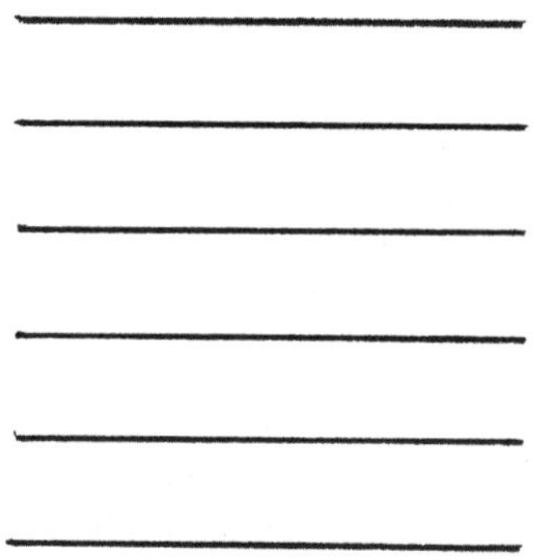

from left to right 6 4,47%

from right to left 3 2,23%

The lines are not pointing in any direction 124 92,53%

1 0,74%

134 = 100%

In which direction are the 6 lines pointing?

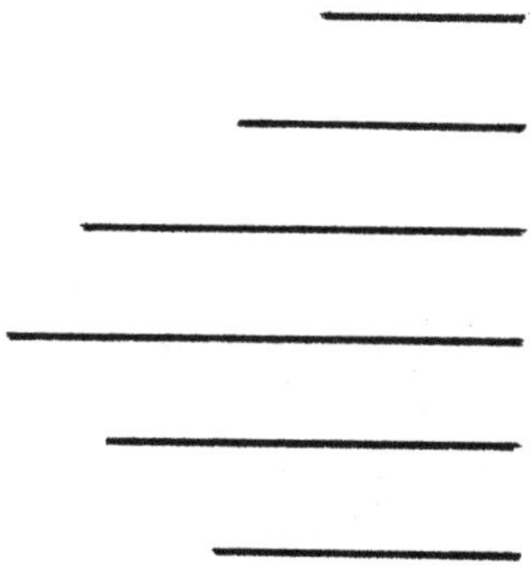

from left to right 18 13,43%

from right to left 115 85,82%

The lines are not pointing in any direction 0 0,00%

NA 1 0,74%

134 = 100%

In which direction are the 6 lines pointing?

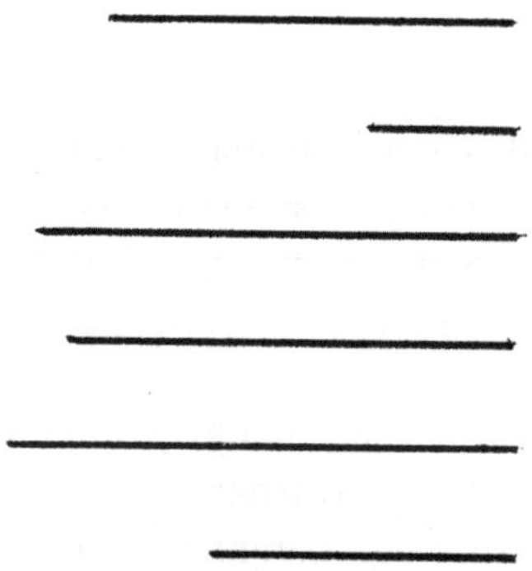

from left to right 36 26,86%

from right to left 64 47,76%

The lines are not pointing in any direction 33 24,62%

NA. 1 0,74%

134 = 100%

11.11.1994

almost 100% saw movement. 13% said it was from the uneven, arrow-like end to the vertical side – in the reading direction – while 85.8% said, in the direction of the arrow-like, uneven ends, that is, from right to left, against the reading direction. In the case of the third figure, with its fortuitously assembled lines of different length, 74.5% still said there was movement; 26.86% from the irregular side to the vertical one – in the reading direction – and 47.6% from right to left, following the variously long lines. 24.6% saw no direction at all.

So once again we see – even in the case of not very typical figures – that we understand a deviation from a simple figure of equally long parallels as movement, or at least as having direction. Apart from this, the unequally long lines arranged to resemble an arrow show that a triangle is read in the direction of the tip lying opposite the vertical or horizontal. Cf. page 100 of my "Gestalt, Bewegung, Farbe"[28] (1975).

Now to return to our fir-tree, to the photograph of it (132) and the information it gives us. We see a tree growing upward, its branches, like rays, stretching out to all sides and in the lower portion bending towards the ground in an almost ceremonial movement. It is the fir-tree's superb gesture that enables us to take note of the space around it.

Even if we do not usually realize why we see how the fir has grown and will continue to grow, we nevertheless do see it. It is that which shows us its individual character. The same obvious process leads to another expression in the case of the thinner fir-tree in the background of the same photograph.

If we also ignore its material, its colour and light-values, there are essentially six categories, already known to us, which, applied to this fir, give us the information described above:

1. the rotating, changing angles at which the branches grow from the trunk,
2. the spacing, changing from dense to wide and back again,
3. the uneven curvature of the branches growing from the trunk. Here, too growth shows itself in a gradual process: at the top the branches spring almost straight up from the trunk, and as they get steadily heavier they curve unevenly more and more strongly.
4. The increasing size of the branches from top to bottom,
5. the varying length of the main and subsidiary branches,
6. the upward pointing triangular or arrow-like form of the whole tree.

These six categories and their metamorphoses provide us with an important part of the perceptual system with which we decode our visible environment. As soon as we apply these categories to a particular object – for instance, a fir-tree – we realize how close to reality they are, in spite of apparently being abstract. We see the categories simultaneously in the same space, at a glance, so to speak. That is why they determine each other, even though individually they are often ambiguous. Let us reconstruct this game in the case in question:

We cannot make out a direction simply from the roughly axial, symmetrical, rotational movement of the fir's branches with their varying angles and density. It could just as easily be read from bottom to top as vice versa. But when we add the tree's triangular form, pointing upwards like an arrow, the rotation automatically appears to be going from top to bottom, it becomes a counter-movement to the upward direction. The tree grows upwards in the direction of an arrow and from there the branches sink downwards with increasing weight. Simultaneously they curve more and more towards the light, to counter the downward movement. This movement is strengthened by the branches' gradual increase in size from top to bottom, in connection with the alteration in density. Apart from this, the impression of vital growth is increased by the varying growth of the variously long branches, which, beginning at the trunk, can only be read as directed outward.

The two most important conclusions to be drawn from the results of all our tests so far are: we interpret deviations by the complex natural forms from their corresponding simpler schematic structures as the result of growth and other processes; these movements make us see the forms as three-dimensional. By comparison with the probably two-dimensional, geometrical perceptual images, we realize which processes have given rise

[28] Jürgen Weber "Gestalt, Bewegung, Farbe", op. cit.

to the natural forms. In this way we read a part of their past, their fate, and so can draw conclusions as to their future.

Thus we understand deviations from the simple geometrical forms as the result of processes deforming a form. They tell us something about the past, present and future of the objects seen, and thus about their fate. They thus have an effect which is three-dimensional and full of expression, apart from which they then become units of a higher and the highest order.

Of course, this does not mean that the expression and assertion of a natural form grows in proportion to its difference from our perceptual cortex's Euclidean concepts. In a test on 21 April 1995 to check the result of the test with 159 participants on 4 November 1994 concerning point symmetrical radial figures (Questionnaire V), in addition to the well-known radial figure we depicted, under D, a figure with three variously wide expansions and contractions, i.e. with different spacings between the radials, but not gradual ones, as in the case of most of the figures in this series, but sudden, more or less fortuitous. Although 12% put a cross against "stiff and unmoving", 88% saw various forms of circular movement: (revolving slowly, revolving faster, revolving connectedly, revolving disconnectedly, revolving and stopping suddenly). By far the largest number of crosses, 77%, was put against "revolving disconnectedly". Obviously the participants recognized figure D as a symbol for an irregular, fast, possibly frequently interrupted, disconnected revolving of the radials. But such incalculable processes do not permit conclusions to be drawn as to past and future happenings. They certainly offer less information about their past and future behaviour than do gradual alterations.

This becomes very clear when we compare a ceramic vessel whose walls happen to be crumpled (62) with one whose walls are gradually deformed (61 a, b). The gradual deformation is much more expressive and expansive than the much more complicated walls of the crumpled vessel, which, since it provides less information, is again a unit of a lower order. It can be made any length, its swellings can be changed around, and still nothing basic would have been altered and it could just as well be lain horizontally.

So now, on the basis of depictions of firs and palms, drawn by the students firstly, just from memory and secondly, after watching three slides (including the fern-leaf), also from memory, we have come to the conclusion that our "reproduction memory" – the short-term as well as the long-term memory in fact – stores the objects in our environment – immediately as a rule – in a strongly simplified geometrical form. By reproduction memory we mean the memory that can imagine and reproduce in some way known objects from our environment when they are not in front of us, for instance, we can draw, paint or describe them. Needless to say, our students would never mistake their own reproductions for real fir-trees – as confirmed by our checking with the various natural-looking firs drawn on Questionnaire XIV. If we were to use wood and needles to build a model wood in accordance with these drawings, it would deceive nobody. Everyone knows that fir-trees look different. If we placed them side by side we would immediately recognize any unnatural simplification – for instance, a strongly geometrical one. We obviously possess two different image-memories.

One, the reproduction memory, contains, without confrontation with the imagined object, various much simplified geometrical schemata – this applies also to colour and movement, as already proven in "Gestalt, Bewegung, Farbe"[29]. The other memory, which we shall call "identification memory", does not come into action until we are confronted with an object. I see a fir-tree, a palm-tree or the face of a close friend and recognize it. With only the help of my imagination, without a confrontation, I could reproduce it only in various simple geometrical schemata. But as soon as there is an optical confrontation, this other memory seems to function in such detail, that I can even see if my friend is happy or sad, ill or well. I see that something has changed, even if I do not know exactly what. A good friend suddenly looks different, but I do not realize at once that she has changed her hairstyle, and often I cannot describe exactly how she used to do her hair, because our reproduction memory is not very good at that.

But have we perhaps artificially exaggerated the contrast between reproduction memory and identification memory by asking our students to draw? Was that after all the wrong method?

[29] Jürgen Weber "Gestalt, Bewegung, Farbe", op. cit.

In February 1993 we showed 316 engineering students the slide of the same fir-tree for 30 seconds. Then we divided the students into two roughly equal groups and set before them two different questionnaires. I have already reported this. Questionnaire A had only very general questions, while B contained eight rather detailed questions.

If we ask questions touching the content of the visual memory, although we are not dependent on the students' artistic ability the questions – especially if they are very exact and detailed – direct their attention to particular qualities of the object in question. The first question was the same for both groups. We asked which geometrical form corresponded most exactly to the form of the fir-tree. As I have already reported, 57.5% decided on the two-dimensional form of the triangle. The second question was "What are the spaces between the branches of the fir-tree?" and the third, "At what angle do the branches grow out of the fir-tree's trunk?" When the students had to draw a fir-tree from memory I had explained that I wanted a section from top to bottom so we could see how the branches grew out of the trunk. But I had made no mention of the spaces between the branches, because this question would automatically be answered by the drawings. In the case of the 1993 questionnaire we had to mention the spacing problem if we wanted to learn anything about it. In the event, only 70% of the students answered the two questions with the simplest, or nearly simplest schema of parallel branches with equal angles and spaces. 15.6% decided that the spaces between the branches were "sometimes large, sometimes small" and 11.3% that they altered gradually. But these students also chose the same angle for the branches; all parallel, either upward, downward or horizontal. Only slightly more than 3% remembered that the angle of the branches rotates from top to bottom, even though we had shown the slide for 30 seconds immediately beforehand. 97% persisted in describing the branches as parallel, so the geometrically simplified result was much the same as the 1998 drawings.

But Questionnaire B produced significantly different results. The second question was, "Are the spaces between the fir's branches:
more or less equal,
alternately large and small,
gradually changing?"
The next question was, "Do the branches grow from the fir's trunk:
horizontally,
upward,
downward,
or all of these?"

Oddly enough, once again 92% – hardly fewer than in the case of the drawings and the previous questionnaire – chose the same angle and thereby parallel branches. But they made better use of our question concerning the different variations of spacing between the branches. Only 39.4% decided that spaces and angles were both equal. 17.5% decided the spaces were "alternately large and small" and 35.5%, "changing gradually". Obviously, the more exact the questions, the more differentiated the answers. But do these questions really extract more from the visual memory, or are we simply offering the students a palette from which they make fortuitous choices or calculate guesstimates. The question "changing gradually" is so exact that it probably radiated conviction, thus accounting, in my opinion, for the high percentage of 35.5%. Despite this, 57.1% opted for "equally large" or "alternately large and small", i.e. for fixity or coincidence instead of a growth process.

I have been unable to determine what exactly the questionnaire in Günther Kebeck's book "Wahrnehmung" (Perception) hoped to find out about the visual memory. I quote: "Please, imagine a bunch of white rosebuds lying among fern-leaves in a box in a florist's shop. A) Are the colours white (green) fairly clear in your imagination? B) Do you see the flowers in a clear light? Is the picture as bright as the objects would really look if they were lying on a table in front of you? C) Are the flowers, the leaves and the box easy to recognize and clearly defined? Can you see the whole group of objects simultaneously or is always only one section clear while the rest is out of focus? D) Can you experience the scent of the rosebuds and the fern and the smell of damp cardboard? E) Can you feel the velvety surface of the petals, the rough surface of the fern, the stiffness of the cardboard?" etc.

Yes, well all one can say is "yes" (to misquote Elisabeth Hauptmann, alias Brecht). The student would not even be lying if he said he could imagine all that. Why should he think that he could not? Anyone can imagine white and green! But if he could really "imagine" it he would have to see before his inner eye that the rosebuds' "white" turns green on the outer petals, while the inside petals near the seed-vessel are yellowish. The same applies to all the other questions. He scarcely realizes how vague his remembered image is. Only detailed questioning can reveal this, which means we are back where we started. Probably one of the best methods to discover imaginative ability and the content of the reproduction memory is to ask the right questions when setting a drawing test – after all, from their second to their tenth/twelfth year children use drawing and painting to develop their reproduction memory.

Certainly our 1993 questionnaires basically attained the same results as the drawings, if not quite so clearly. The remembered image without a visual confrontation, our reproduction memory's knowledge, is geometrically simplified. This also applies, in a similar way, to colour and movement (see "Gestalt, Bewegung Farbe"[30]).

However we imagine the construction of the visual cortex, we have now discovered fairly precisely two ways in which it functions: firstly, we *compare* environmental phenomena with the simple geometrical form most closely resembling them, from screen to circle, and the schemata composed of them. We recognize deviations to be the result of processes, for instance, growth. We see forms which we realize have been deformed by a process as three-dimensional. We have tested this on models with uneven curvature, increasing and decreasing density, various size and length and different angles. We then applied these models to living firs and palms to explain why we realize that these trees are growing and why these visual processes make the trees appear more three-dimensional than a corresponding three-dimensional geometrical body. This was demonstrated particularly well by the schematically drawn fir-tree C (Questionnaire XIV). We drew the branches and needles or twigs with rotating angles from upward, through horizontal to downward. We depicted the branches as unevenly curved. Although we did not use any spatial methods of depiction in this drawing (perspective, intersection etc.) only 19.3% of the students described the fir-tree as "flat", while 80.7% said "three-dimensional". Looking at the simplest geometric schema, on the other hand, 98.4% said "fixed", "unlively" and "flat". This was clear proof that if an environmental figure completely corresponds with our visual cortex's schematism of simplicity it is seen as fixed, unlively and flat, giving little information or expression.

But in order to reproduce the schemata of our environment's forms our reproduction memory uses precisely our visual cortex's simple perceptual images, with which we compare reality in order to arrive at conclusions. Our reproduction memory organizes them and can recall them. The perceptual images become foundation on which we build the memory-pictures of our environment for our reproduction memory. The very differences of the visual world to this form-schemata allow us to recognize the processes by which the objects in our environment are made and in so doing we also recognize their expression, see figures 361 and 361 a, b and c. Themore visual experience we gather, the more differentiated and complicated become our reproduction memory's form-schemata and thus our perceptual power of judgement. This is a further development of our visual memory step by step and hence of our perceptual ability to think. This phenomenological "top-down" or "bottom-up" process is probably equivalent to the corresponding neurophysiological images. More on this later.[31] Almost all the so-called intellectuals do not get any further than the level of 10 – 12 year olds in our current schooling and educational system. This means that their reproduction memory knows only those schema consisting of the simplest geometric forms.

At the moment I still do not know why our identification memory, which cannot become active until there is visual confrontation, appears to store memories so much more precisely, nor how it does so. Perhaps it is in some way connected with "priming". Animals also seem to have such an identification memory, while there seems to be only slight evidence for the existence of the beginnings of a reproduction memory. I refer you again to the experiment with the pigs and the square and the well-known experiments with chimpanzees (Köhler) or bees' dances (Frisch).

However, I do not believe that we should think of the identification memory as a sort of photographic memory that can be called upon when there is visual confrontation. It may well be very

[30] Jürgen Weber "Gestalt, Bewegung, Farbe", op. cit.

[31] A. K. Engel, P. Fries, W. Singer in "Nature Reviews/Neuroscience", Volume 2, October 2001

like the reproduction memory, but enriched with typical details: a friendly or twisted smile, a particular way of moving the mouth, such striking details as eyebrows or moustache, a very impressively shaped head and so on. Perhaps the identification memory, basically schematic but enriched with such characteristic details, is able to recognize the individual and not just the species. This calls for intelligent experiments. It is a fact that although we recognize an old friend by means of our identification memory, we are usually still surprised by many details we had forgotten. It seems that the identification memory is not, after all, so very precise. The reproduction memory is certainly of crucial importance for our consciousness. Neglecting its tuition therefore proves terrible ignorance of the development of human thought.

I am assuming that in the case of infants the development of the ability to recognize objects they see – i.e. the existence of the identification memory – is not concluded before the beginning of the second year, at about the same time that the creation of a reproduction memory manifests itself in drawings which resemble each other all over the world. It is true that a baby recognizes its mother in its third month. However, even those who are not engaged in research into the behaviour of babies realize that a seven or eight-month-old baby forgets things it has seen sooner than an adult does. For instance, if a well-known relative is absent for a few months, the baby has to get to know him or her all over again. This can be observed from the long, intensive stare to which new acquaintances are subjected, and which only stops once the person is familiar, but which is repeated if the person is absent for some time. For instance, granny and grandpa are not spontaneously recognized until some time during the second year, at the time when the reproduction memory is being developed.

Chapter 9
The "Orbits" and Their Application

Of course, it could be argued that such simple patterns as the Euclidean geometrical forms or the Lie "orbits" were unsuited to decode the expression of the complicated and extremely manifold phenomena of our environment; to understand the processes creating them; to recognize the essence of things by comparing them with geometrical forms. That is certainly the case, and that is why systematic perception schooling is essential both for children and adults – as we have already ascertained – to create increasingly complex schemata. But it is nevertheless astonishing what conclusions we can draw just with these relatively simple patterns which, somehow or other, are inscribed in our brain.

Thus it is that the Lie Transformation Group's "orbit" and its metamorphoses, organized by means of even radials, is an important key in the entire botanical world for the expression of three-dimensional growth. We have already used it for several experiments with students, concerning radial expansion and contraction, extension and diminishment, and we have also used it to decode the principle of growth in firs and palms.

An example of this is the willow (174). At a height of about two metres it divides into three or four variously long radials, drifting variously wide apart. New centres of growth then spring up from these, from which, on the right of the picture, variously dense and unevenly curved new branch-radials diverge. These then continue to form new radial centres up to the treetop, ending finally in the long, thin, downward swooping branches of the weeping willow. This expression is strengthened by the uneven curvature of the branches, most of which are read from the flat to the strong vaulting. This expresses the dramatic struggle between the tree's growth and its expansion in all three dimensions with the prevailing north-west wind. These swoops of the uneven curvatures from bottom left to top right are spatially ended by the equally vaulted, variously dense, downward hanging twigs; this time proceeding from strong to weak curvature. The three-dimensional, dynamic effect is now based on the variously dense parallels of the downward hanging twigs which, proceeding from the other "orbit", we have already tried out on students (Questionnaire VII). The dynamic is also increased by the enlargement of the branches' circumference from north-west to south-east; the main trunk divides in a radial manner at a height of some two metres. But we shall have to examine separately the expression associated with altering sizes.

In practice every tree, every bush, every plant is an example of what has just been said (114).

We have now explained the expression of spatially moving growth by comparing unevenly with evenly dense radials, unevenly with evenly dense parallels, uneven with even curvature and uneven with even size. Of course, we have not yet included in our observation the tree's rough or smooth bark, or its colour. Nevertheless it will be difficult to deny that the expression of the three-dimensional appearance of this tree and other plants can be explained by comparison with four relatively simple geometrical categories. These are also partially able to decode the manifold expressions of growth in the great majority of this world's plants, however different their appearance.

Let us now use these systems to examine two landscapes.

Even though light values and intersections contribute essentially to the impression of space, we cannot ignore the fact that in one landscape (111) the contraction and expansion of the radials, as expressed by the trees' growing apart beside the path, considerably increase the dynamic of the three-dimensionality. Of course, the deep shadows on the ground also contribute, but even they are organized in variously dense radial sections. Their centre is in the picture's right foreground, while the centre of the radially diverging trunks is below the picture's right outer edge. Compared with the dramatic appearance of the trees lining the path between two pools and so diverging either side of the water, the simple view into a wood – a space containing the alternating density of the mostly vertically directed parallel trees – seems relatively peaceful (112). But here too, apart from the mutual intersection, the spatiality is much increased by the gradual process from contraction on the right to expansion on the left, and then contraction again on the right outer edge of the picture. In both pictures, movement is strengthened by the occasional uneven curvature of the trunks. As I have already said, that alone does not explain the dramatic or peaceful spatial effect of the two landscapes, but there is no doubt that the alternating density of the parallels or the radially organized trunks and shadows contribute an important spatial and dynamic structure to these landscapes.

Koffka and Köhler, and in the late sixties Arnheim, started from the principle that the recognition of an object's whole simple geometrical form corresponded to the geometrical constellations in the appropriate parts of the brain. Arnheim emphasizes these thoughts in his book "Art and Visual Perception" written in the fifties, and concludes that we do not recognize the environment until it conforms with these perceptual images in our brain – this is what he calls the geometrical ideas. Arnheim said that, "if we leave the world of well-defined, man-made shapes and look around in a landscape, what do we see? A mass of trees and brushwood is a rather chaotic sight. Some of the tree trunks and branches may show definite direction, to which the eyes can cling, and the whole of a tree or bush may often present a fairly comprehensible sphere or cone shape... And only to the extent to which the confused panorama can be seen as a configuration of clear-cut directions, sizes, geometric shapes, colours, can it be said that it is actually perceived."

Similar statements came from Hoffman and others, who thought they had made out their "orbits" as important structures in our brains. Hoffman, Dodwell and others consider perception to be a conformity between these "orbits" and what is seen, so at least I must think, on the basis of Hoffman's writing on "Op art". I know that it was really so for Arnheim – we discussed it for fifteen years. He was convinced that each geometrical form, whether circle, square or rectangle, was a dynamic image, and truly understood perception could not occur until there was conformity between it and what was seen. He never accepted my idea, as presented in my book "Gestalt, Bewegung, Farbe" 1975, 3rd edition 1984: that we perceive only by means of "comparison" with these geometrical concepts and that it is the deviations between them and the natural forms that let us recognize the dynamic of what we perceive; and that contrary to his ideas, complete conformity with our brain's geometrical ideas leads to a lack of information. Many discussions, carried out over many hours, changed nothing.

But it seems to me that our experiments to date and also the analyzes of natural phenomena confirm, on the one hand, the existence of such geometrical ideas and, on the other hand, that the actual process of perception is the comparison with these geometrical concepts and that we gain our visual information from the deviations from them – not from conformity with them. In this light, the Lie Group's "orbits" and Arnheim's less precisely defined geometrical structures, are good starting points for understanding the essence of perception. Engel, Fries and Singer probably mean the same thing when they say that these models require the comparison of inputs with existing knowledge as essential for perception[32].

But the "orbit" deformed to unevenly dense radials or parallels is not just a key to the expression of the botanical world: let us consider the face of my "Dionysos", dating from the seventies (175). It is quite obvious that the bridge of the nose between the eyes is the centre for unevenly dense revolving radials, depicted by the nose itself, then by the deep, negative lines dividing the cheeks from the upper lip and the bags under his eyes, and also by the eyebrows and forehead vaultings, directed at this centre in five different angles, and by the lines edging the mouth, which, if lengthened, would also arrive at this centre. These revolving radials give the face a wild, dynamic expression, extremely different from the peaceful and relaxed one, built on a horizontal-vertical system with differ-ent spacings, of the Cambodian Buddha (176). The face's gentle, mostly upward directed unevenly curved vaultings increase the expression of blooming, happy youthfulness.

In complete contrast is the Buddhist world guardian (177), dating from the ninth to tenth century. The uneven radials, also directed at the bridge of the nose, together with the powerful, upward directed vaultings of the mouth and cheeks increase the wild expression of the head, although quite differently than in the case of Dionysos. That is because in his case the unevenly diverging radials unite less with full, upward curved vaultings than with surfaces which collide relatively sharply and rotate. We shall consider later what is meant by rotated surfaces – another perceptual category. The wild expression of the actor's mask from Tarento (fourth century BC) is again completely based on unevenly dense and variously long radials. The radials are the eyebrows, the upper and lower lids of the wide-open eyes, the bags beneath the eyes and the creases in the cheeks (178).

Thus we see that with the alterations or deformations of the radial "orbit" or of the "orbit" with the horizontals and verticals, applied to the various objects around us, we have found a deforming category, but also one that can be detached from our environment, we can discern expression and information from its variation from the simple geometrical "orbit". Of course, this does not explain the facial expression because, for instance, we have not considered colour and surface quality. Despite this, drawing on the three categories, we can differentiate expression and information very clearly: the comparison between uneven and even vaulting, uneven and even spacing, between radials and parallels and rotated surfaces compared to parallel surfaces.

At this point I can hear some readers saying: No, the very important question of the position of the mouth has been left out of the facial analysis. What a difference there is in a facial expression depending on whether the corners of the mouth droop in resigned depression or are bent upward in a smile. Correct, but such a detail must not be examined out of context. Let me return to the picture of the two old men, the ship owner and his foreman (59). The corners of the foreman's mouth are turned down, the ship owner's, up. Try it yourself, in front of a mirror. The moment you turn the corners of your mouth down, you also pull down the curves of your cheeks. The entire vaulting of your face becomes flatter and is directed downward. If the corners of your mouth curve upward, cheeks, palate and the surroundings of your eyes also curve upward. Even your nostrils are affected, pulled downward or upward, as the case may be. The various movements of the mouth have an enormous effect on the vaultings of the face, which have the greatest influence on facial expression. Although the Buddhist world guardian's mouth droops downward (177), his vaultings are directed upward – a sculptor can do this – so his expression is not in the least depressed; on the contrary, it is aggressive and vital. In observing a natural face we also realize that the deformations of the facial vaultings are the most important factor in the expressions created by the movements of the mouth. It makes no difference if the faces are young or old. The movements of the mouth alter a natural face's vaultings from the eyes to the palate and that is the decisive factor in altering the expression.

Even if our brain did contain a centre specially for identifying faces[33], we still read facial expressions in accordance with the same categories we use for everything else – as the examples quoted so far

[32] Engel, Fries and Singer, in "Nature Reviews/Neurosciena", op. cit.
[33] Francis Crick "The Astonishing Hypothesis: The Scientific Search for the Soul", op. cit.

prove. Of course there is a difference between judging a fruit's or a face's directed vaulting; uneven radials in a plant or facial lines. The same categories, applied to different objects, have similar but by no means the same effect on expression.

As far as the reproduction memory is concerned, faces and other objects are one and the same.

In the first lecture of the summer semester 2000 I asked each student to take an A4 sheet and draw a full-face and profile portrait of his or her father – a person whose features were very familiar to them. The unsurprising result was that apparently 90 students – with only three exceptions – possessed only two different fathers: one bearded, the other clean-shaven. In profile he was reduced to one type, constructed with rather flat, straight lines. We shall examine this more closely later.

At this point the gentle reader will again blame this rather on lack of drawing ability than poor memory. So I urge you to try an experiment I could hardly carry out with a hundred or more students: ask your husband, or wife, or one of your children if your own upper lip, seen in profile, is in front of, behind or level with the line of your forehead. Of course, they must do this without looking at you. To your amazement you will discover that correct answers remain at the chance quota of 33% at the most. The forehead and upper lip were almost always level in the students' drawings, because this is the simplest form.

As I have already said, our reproduction memory depends on our brain's geometrical ideas and does not work photographically. This assumption may appear speculative, but our experiments to date all point in that direction, as does the beginning of children's drawing and as do the tests to discover what babies recognize first. And there is a still more amazing proof.

Chapter 10
The Start of Ornamentation All over the World and at All Times

On all five continents ornamentation on household objects, but frequently also on depictions of figures or animals, starts with the literal use of orbit-like geometry, or, to use a more general expression, with the same geometrical figures we see in children's drawing after the scribbling stage. All over the world ornamentation starts by reproducing the geometrical perceptual images which are then altered more or less quickly by observation of nature or by variously dynamizing the geometrical forms. In some places, for instance in Greece, this happened in the course of a few centuries, while in other countries, for instance in Australia, it has stayed the same, from the drawings on rocks made some tens of thousands of years ago, to today. Let me draw your attention to the parallels ornamenting the seated man in an Australian rock drawing, 10000 BC, (179) and a corresponding painting on bark (180) from recent times, and to the contemporary parallel hatchings on the bark painting with the inset crocodiles (181) similar on Taiwan (184); everywhere the same use of parallels for ornamentation. But the circular "orbit" with radials is also frequently found in the South Seas: as a pattern on material (182), decorating an old ceremonial paddle (183) or on large decorative objects (187). We also find the spiral ornamenting old wooden figures or shrouds (185, 186).

Even if, understandably, the catalogue on African art published in Berlin in 1996 describes the objects it depicts as unique in the world, it cannot be denied that not only the geometrical design of the figures but also the ornamentation have their roots in the same perceptual presuppositions. Even if the authors of the fourth chapter, "Central Africa", Daniel Biebuyck and Frank Herrmann, claim that the material painted by the women foragers in Zaire is spontaneous and unique in its artistry, it is still clear that the vertical and horizontal parallels, the screen-like crossings (199) and the radial figures correspond to the classical "orbits" found all over the world at the beginning of the development of ornamentation. Of course this African barkcloth is probably only a few centuries old, but above the parallels we see a pattern of radials crossing in the centre, which look like some of the children's drawings or the oeil-de-boeuf windows of the patient with Alzheimer's disease we have reproduced and which, as we shall see, reappear on every continent. They are not, of course, the result of cultural imports or exports, but neither are they the unique invention of African women artists (this barkcloth was painted by women). They are simply slightly individual variations of the same mental perceptual images common to all humanity, without regard for race, origin and trade routes.

Even the lines scratched in a stone in South Africa (198) sometime between 1000 and 2000 BC correspond not only to Hoffman's "orbits" but also to children's drawings. As long as ornamental styles and cultures keep to this common Euclidean geometry, it is almost impossible to make out really individual differences between races, nations and continents. Of course, on the basis of certain variations, it is possible to differentiate between African and Australian bark paintings. The bases of the ornamentation – bark or vases – the material of the household goods, all make for differences; but these should not be overrated. Everything is based on common human perceptual images. These are not individually diversified and used for different motifs and at different degrees of development until a high cultural level is attained.

On the continents the origins of ornamentation were staggered throughout the millennia, because some – for instance America – were settled much later than others. But the correspondence between the beginnings on all the continents, which may have started thousands of years and kilometres apart, are so amazing that they cannot be explained by the art historians' usual assumption of the export and imitation of forms. They have to have an almost exclusively mental base. Everywhere, and completely independently, it must have started thus:

Let us begin with relatively early examples, with the start of culture in Mesopotamia, Iran, Industan and Egypt. Admittedly, we have here a block of geographical neighbours. Correspondence here could be the result of cultural export and imitation, were it not for the fact that very similar ornamental geometrical structures are to be found from Japan, to Australia to America at completely different times. On a crater from Persia, Tepe-Sialk, c. 3200 BC (188) (from the Hetjens Museum, Düsseldorf) we can see all these early ornamental forms, just as if it were a pattern book. As I said, I am including here not only the Lie Group's "orbits" but also children's drawings. Ignoring the animal depictions at the top, we find next parallel lines. The Lie "orbits" of group B consist of these and we also find them over and over again in the scribbles children produce, though less precisely, in their second and third year. Going downward we next see the wavy lines so characteristic of the next stage of children's drawing (18) and in the main ornamental band we see, on white rectangles, the circles surrounded by radials which we find in children's drawings shortly before they invent the tadpole (34, 35). On this vessel they may be intended to symbolize the sun (but that does not matter, as we shall see) and between them is the chessboard pattern which is the first to be recognized by babies and which turns up time and time again as a simple screen-like crossing pattern in children's drawings. Of course, what we see here is extremely disciplined, with a highly decorative effect, but there is no denying that these are still the same geometrical forms. Triangles are there, too, every one underlined by three parallel lines. Figures (189–194) provide further examples of the chessboard pattern and parallel lines, also from the same era in Iran, zigzag lines and linear screens, parallel hatching dating from between 3500 BC and 1000 BC, circles inside circles surrounded by radials.

The proto-Hittite stag, 2300 BC, is ornamented with orbit-like circles within circles, zigzag patterns and the cross which also turns up in children's drawings in their third year (195, 18, 24).

The Afghan vessel from Ulug-Tepe, third millennium BC (203), is also decorated with the zigzag line, strongly reminiscent of the ornament on a North American vessel dating from between 1100 and 1300 AD (219). The bear from the Cyclades, carrying a vessel, 2000 BC (196), has a screen painted all over his back and parallel lines on his front. Screen patterns seem to have been used to represent pelts or hair for a very long time, for instance: the hair on the head of the ivory statuette from Landes, France, 25000 BC, (200). The archaic Cypriot figure of a woman, 700 BC (197), was still making use of a similar screen to represent the hair.

The middle Cypriot vessel, 1700 BC, is decorated with diagonal screens made of stripes (201). It is amazingly reminiscent of the Central American vessel from the Chancay culture, 1000 AD, painted on one half with a chessboard pattern and on the other with a horizontal, vertical and diagonal screen (217) (Hetjens-Museum, Düsseldorf). The Afghan vessel (204), c. 3000 BC, has an inexact chessboard pattern on its neck and a very interesting combination of rectangles and horizontal/vertical crosses on its shoulder like the ones found individually in the children's drawings (24, 31). Both geometric designs are centred one over the other here, typical of Afghanistan at this time.

Similar ornaments, separated by continents and millennia, are always found at the beginning of a culture. It would be misleading to consider cultural export or imitation to be the reason for this. The invention of forms started everywhere with the reproduction of the geometrical perceptual images; further development then followed, slowly or quickly.

The Attic crater (226) dating from the ninth century BC like the hydria (Troy VIII) from 700 BC (225) are interesting examples of the development of the meander from a simple geometrical form, the right angle. Of course, it is possible to understand the red or the yellow backgrounds as meanders, but only the angles in between are outlined, and are thus figures on a red or a yellow background. This is amazingly reminiscent of the child's drawing (30). Even though its figures are not opposite each other, as on the Attic crater, it has the same angles. Admittedly they are parallel, but they, too, are organized in a screen. Even simpler is the pre-form with rod-like shapes, which structure the background using meanders on the shoulder of the geometrical amphora (223) or the native American vessel, 1500 AD (224). The crater from the eighth century BC (227) demonstrates a further step towards the meander: a rectangle filled diagonally with step-like lines, reminiscent of similar diagonal step-like lines on a Central American censer dating from the first 500 years AD (218). Here, too, at the bottom we find a true meander, like the one on the eighth-century BC Greek amphora (228), yet more complicated is the one dating from 330 BC (229). From the amphora (223) via the Attic crater (226) to the amphora (228) we can follow the step by step development of the meander from the parallel "orbit". This is thus an example of figure and background changing in the development of an ornament. I would like to remind you of the gestalt psychologists' picture that changes from a vase to two profiles. The amphora (228) is otherwise mainly ornamented with horizontal and vertical parallel lines, while in the area of the handles, on square fields, we find absolutely classical forms of the two A "orbits" with circles and radials.

The development of the spiral can also be followed step-by-step. The spiral, as an ornament, develops from the circular "orbit". Hoffman's "orbits" on the tiara of Ramses II (215), or the geometrical amphora (206) are transformed on the Attic cup, eleventh-century BC (209) and the pyxis from Siros, 3000 BC (210). They show the development of the spiral from the circular "orbit" through diagonal lines. A very interesting intermediate form between the circular "orbit" and the spiral is to be seen on the Mycenaean-Trojan vessel (1700 – 1200 BC) (207), where several branches have turned the circular "orbits" into a spiral-like figure.

We find a whole series of really free spirals, covering the entire surface (they are said to represent running water) on the Cyclades pan, c. 1000 BC, (205). If we then add the spirals engraved in the passage tomb to the north of Dublin, 2500 BC, (234, 235) and remind ourselves that the Greeks had used pre-forms of the spiral between 3000 and 1700 and 1100 BC, we begin to realize how pointless it is of archaeologists and art historians to wonder where the spiral actually came from and how it spread throughout the world. These geometrical figures have to be of mental origin and were newly invented independently in many places in the world based on the circular "orbit". And this is without taking into consideration that on the Cyclades pan the spirals meant water - there is a ship[34] between the spirals - but in Dublin they meant the sun.

Again and again we find the same geometrical forms, whose origins lie in the human brain's visual cortical areas and whose meaning varies from place to place: in the royal Egyptian necklace, 2550 BC, (212) developed from circular "orbit" and external radials; in jewellery (213) from the same era; in a prehistorical Egyptian clay vessel from the Negade culture (Hildesheim) (214), fourth millennium BC, with screens, parallels and triangles; in the fire pillar of beaten gold, thirteenth century BC, from the Palatinate (231), found in this form nowhere else in the whole world, decorated with parallel lines and circular "orbits".

Today we still find the same parallel screens and circular "orbits" on the ceramic vessels of black Africa, for instance in Salah (238), as well as on the centuries-old figure from the Dengese tribe. Its diagonal screens, parallels and circular "orbits" reflect the then prevalent tattooing (237). Whether the comparable ornaments on the anthropo-

[34] Manfred Korfmann "Troia als Drehscheibe des Handels im 2. und 3. vorchristlichen Jahrtausend", in "Traum und Wirklichkeit in Troia". 2001

morphic vessel from Vidra near Bucharest represent tattooing or clothing (240) is irrelevant; all over the world we find the same, very limited collection of geometrical forms. We even find similar beginnings in China (241). This is clearly shown by the vessel with two large spirals from Hsientien in Kansu, second millennium BC and by the vessel with the three spirals from Kansu, also from the second millennium BC (211). The Neolithic Japanese vessel, fourth millennium BC, (244) is decorated with zigzags and circular "orbits". The ornamentation of the Korean belt buckle from the "proto three Empires period", first century AD, looks like a Lie "orbits" pattern book (245). The Japanese figure from the late Jomon period, 5000 – 1000 BC, is decorated with parallel lines and circles, "orbits" and spirals (242). The Syrian vessel, c. 5050 BC, is ornamented with parallels and regular diagonal patterns (216). The Korean helmet, first-century AD, consists almost exclusively of different forms of repeating patterns, which overlap each other artistically (246). The bronze Chinese container, sixteenth – fifteenth-century BC, is essentially decorated with a pattern of burls (247).

Even if the Jews have not bequeathed much fine art – partly because of their culture, but partly because it has been destroyed – there is no denying that the menorah, the seven-branched candelabrum, which is so important to them, is half of Hoffman's "orbit" A with a vertical axis. The Eye of God was a triangle and the Star of David consists of two equilateral triangles. As I have already said, there must be more geometrical perceptual images than Hoffman's "orbits". Obviously there are no technical reasons for these artefacts and symbols (menorah, Eye of God, Star of David). There is just the fact that the unaltered geometrical perceptual image is understood to be particularly basic and universal, simply because it does not say anything individual. If the ornamentation of all the world's primitive cultures since Mesolithic (315) times has been drawn from such a small collection of geometrical forms – the Australian Aborigines retain them even today – this cannot be explained by cultural import and export, especially as most of the examples I have cited – and they are just a small selection – come from eras when long sea voyages were hardly possible. Heyerdahl's many failed attempts to sail from Egypt to America in papyrus boats were based on mistaken art-historical thinking, which believed all artistic forms stem from a single source and subsequently spread out from there. But the source is biological and genetic, based on the structure of the human brain. It has no need of ships. Let us remember in this connection that in eras when voyages on the high seas, though involving much loss of life, were quite common, advanced cultural development was enormously divergent.

However the "orbits" are calculated, I, too, in my book "Gestalt, Bewegung, Farbe", first edition 1975, third edition 1984, started without calculation from the premise (at that time I did not know about the Lie Group's "orbits") that our brain had to have geometrical concepts with which it compared natural forms and that it comprehended the differences between them as the result of processes. Our brain then constructs its form schemata out of the same geometrical forms with which it compares our environmental phenomena and then, via visual objects, retains them in its reproduction memory. In this manner our brain arrives at more complicated geometrical schemata; the next step is to compare these with the more complex environmental forms and come to a perceptual decision on the basis of the differences. This would account for the unexpected neurological diagnosis that visual memory images are stored in the same parts of the visual cortical areas which first receive the corresponding visual environmental images. This is the precondition for seeing and judging becoming a perceptual act. We see natural forms through a sieve of geometrical schemata and judge them through comparison. The whole thing is a "top-down" and "bottom-up" process. Specific experiments will have to clarify whether this has anything to do with the identically named chemical-electrical processes in neurons.[35]

In addition, the fact that certain groups of neurons start to react chemically and electrically at the sight of narrow strip lighting or a sliver of light, makes us suspect that there are groups of neurons which primarily react to very simple geometric phenomena. There are obviously different responses to simple and complicated visual impressions in the various layers of neurons[36].

[35] Engel, Fries, Singer in "Nature Reviews/Neuroscience", op. cit.
[36] Francis Crick "The Astonishing Hypothesis: The Scientific Search for the Soul", op. cit.

This is not sufficient proof as far as neurophysiologists are concerned. Not only the neurons should be examined, however, but also the visual phenomena which our brains create in some way.

It must make us stop and think, when we realize that all over the world infants draw the same geometrical forms, which are the starting point for ornamentation in every culture and which are finally transformed into complicated ornaments such as we find in the classical periods of Greek and Roman art or in eighteenth-century rococo art. These are then reduced to geometrical basic forms again when patients suffer from diseases affecting perception.

The difference between these geometrical ornamental beginnings in world cultures and the geometrical forms in infants' drawings – apart from the differing exactness – is primarily that the geometrical forms in infants' drawings come from movement: up and down, backwards and forwards and rotation. In other words, basic processes, written down by the children – we shall return to this point. On the other hand, the cultures start with these ornamental geometrical forms – there are possibly still traces of movement as the forerunner of geometric Euclidean forms in the "macaroni-forms" of the early ice age. In the course of their development, these geometric forms are distorted by means of imagined movement or geometrically and dynamically interpreted natural examples. The final results are completely different ornamental styles, typical of their region.

Although it is correct to say that form was much influenced in the areas of architecture, sculpture and painting by import and export and by contacts between cultures, art history ought to recognize that there are also biological reasons for the dissemination of very similar art forms. These have less to do with artistic communication than with genetic preconditions. There is also a biology of art.

To avoid misunderstandings, I must return once again to the two analyzed forest landscapes. Of course, the three-dimensional impression is primarily based on the mutual intersections between the tree trunks and branches and the law of the simple form. We also comprehend the path with the diverging tree trunks as three–dimensional because, understood in this way, it has the same breadth along its entire length and the spaces between the – admittedly very different sized – trees are more even (111, 112). The same applies to the view into the wood. There, through three-dimensionality, all the tree trunks are much the same size, so once again this leads to more simplicity. But it cannot be denied that the actual – not just that achieved by perspective – density of the tree trunks on the left and the extreme right and the expansion in the right-hand third contribute considerably to the spatial effect, as we proved by Questionnaire VII. In the same manner the wayside trees between the ponds, diverging like radials at various angles and spacings, are far more dynamic than evenly spaced, parallel-growing trees. They, too, increase the spatial impression, which, as I say, is the result not only of light values, colour and mutual intersections and the law of the simple form, but also – and this is certainly no less important – of the various actual densities of the parallel or radial divergence of the trees. The changing densities, the changes of curvature and negatives, that is the deviations from the schematism of the simplicity of our brain's concepts or from Hoffman's parallel "orbits", create the impression of movement, and thereby of spatiality.

So far we have not differentiated between actual movement and movements in the past, visually rendered by forms. We shall return to this point.

Chapter 11
Actual Enlargement and Reduction

Up till now we have conducted our experiments exclusively with two-dimensional patterns; let us now test them on three-dimensional models. If we set up nine, evenly spaced vertical cubes (249) we recognize a three-dimensional cube comprising nine parts. However, we read it from its surfaces, which are limited at the sides and the top. It is a three-dimensional image that, despite the spaces, seems relatively flat, especially when we compare it with a composition created from nine identical cubes, densely and loosely grouped, so that in the densest parts they even interpenetrate. This group is not only more mobile and interesting, it is also more spatial. In the case of model (248) the spatiality rather lessens the simplifying effect, when compared with model (248) and its screen-like organization, where the three-dimensional effect leads to our seeing not only all the surfaces of the

XVII

Please compare the four compositions consisting of cubes A, B, C, D and decide which description you find the most appropriate for each composition of cubes (A, B, C, D). Please tick the description in the boxes under A, B, C, D, but just choose one description for each composition of cubes.

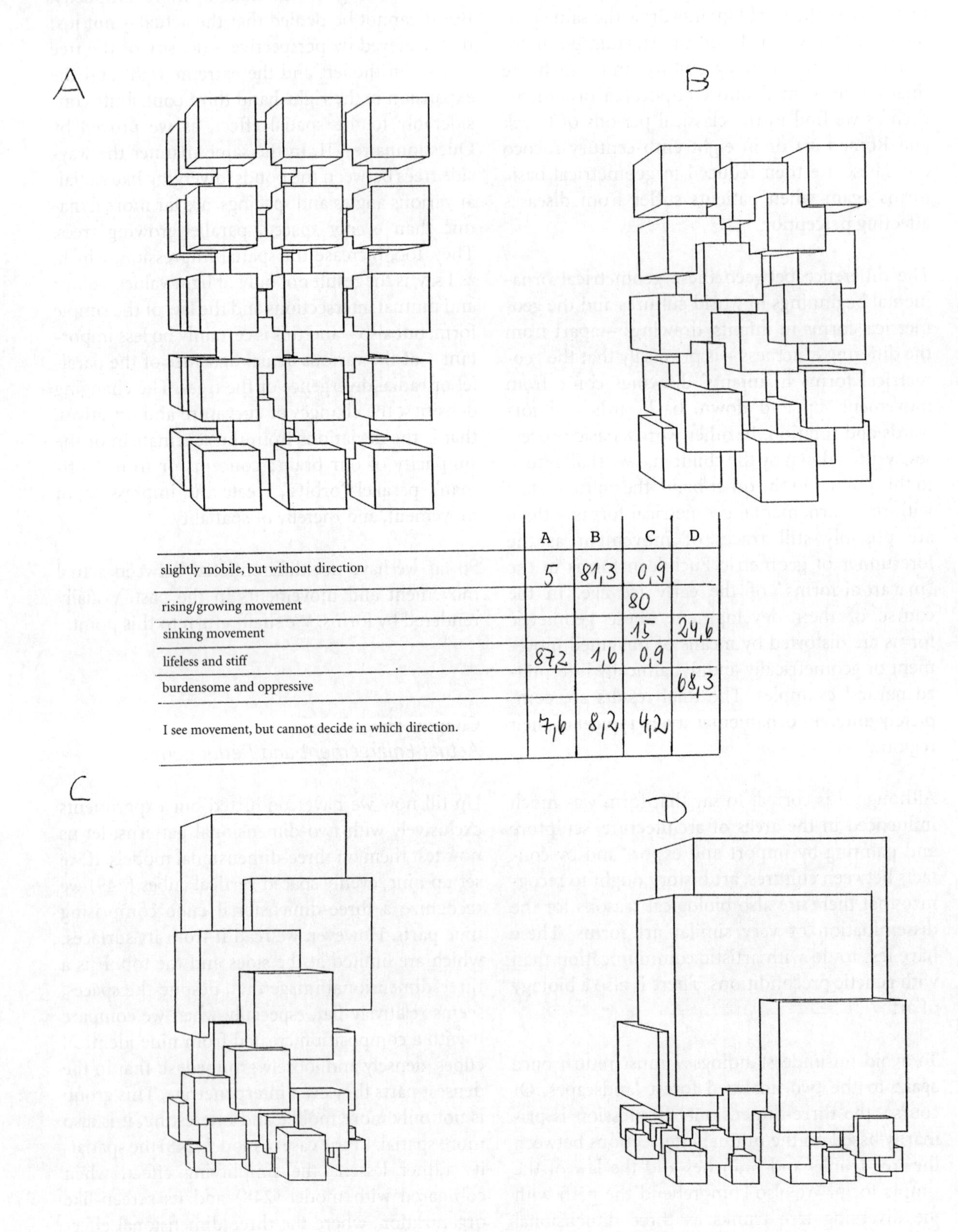

	A	B	C	D
slightly mobile, but without direction	5	81,3	0,9	
rising/growing movement			80	
sinking movement			1,5	24,6
lifeless and stiff	87,2	1,6	0,9	
burdensome and oppressive				68,3
I see movement, but cannot decide in which direction.	7,6	8,2	4,2	

XVIII

Please compare the two compositions consisting of cubes A and B and decide which description you find the most appropriate for each composition of cubes (A and B). Please tick the description in the boxes under A and B, but just choose one description for each composition of cubes.

A

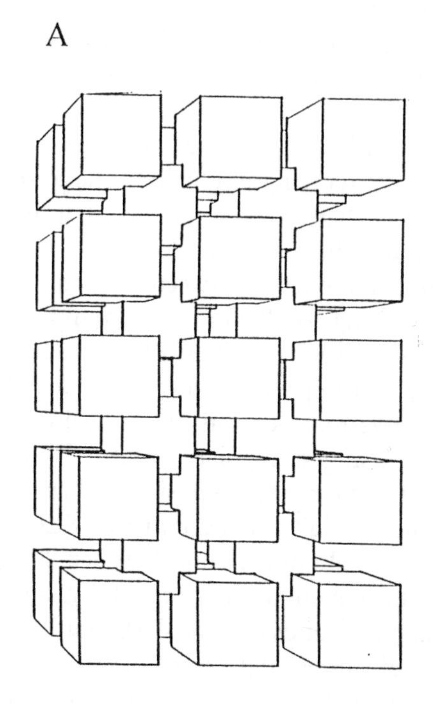

B

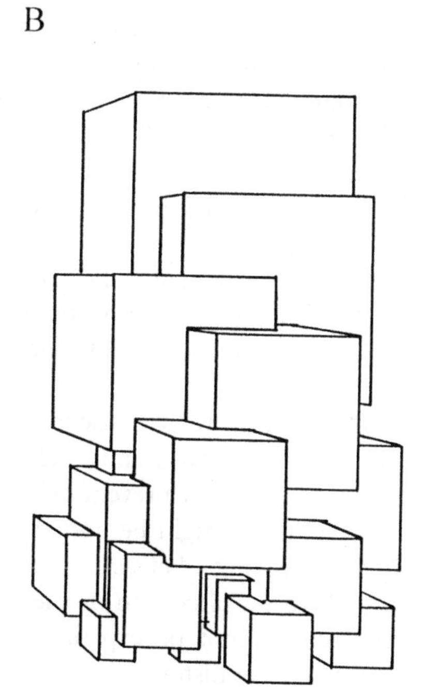

	A		B	
	%	number	%	number
slightly mobile, but without direction	10,2	13	3,2	4
rising/growing movement	1,6	2	51,9	66
sinking movement	1,6	2	30,7	39
lifeless and stiff	81,1	103	3,2	4
I see movement, but cannot decide in which direction.	3,9	5	10,2	13
confused or contradictory information	1,6	2	0,8	1
	100%	127	100%	127

entire cube, but also all the surfaces of the individual cubes and all the spaces between them as absolutely the same.

In the case of (248), spatiality means that only the surfaces limiting the individual cubes are rectangular; the form as a whole and its densities remain complicated and different. Despite this, the effect is more spatial. Unifying the law of the simple form and the deviation from the schematism of simplicity leads to an impression of movement and thereby to more spatiality as a whole. 250demonstrates particularly well to what an extent the process from loose to dense suggests movement. Against all logic we see it as a rising, floating group and are not at all worried by the impossible statics of the situation. The movement seems to bear itself. Many pictorial compositions are based on such phenomena, for instance, Michelangelo's "Last Judgement". The movement of the rising dead is largely based on the phenomenon (252, 253) of increasing upward density.

In some of our models with dense and loose forms we also made use of large and small circles. However, we were unable to prevent completely the possibility that these might be understood as being at different distances. Let us now consider compositions where the alteration of the sizes is definitely real, and cannot be explained as the result of greater or lesser distances.

In 1994 we set the students a questionnaire with four compositions consisting of cubes, i.e. undirected forms. We suggested six possible answers, one answer only to be selected for each composition (Questionnaire XVII). Model A corresponded in another way with the model (249) already described. It is a three-dimensional grid, consisting of 38 identical transposed cubes. 87.2% of 122 future architecture students condemned this composition as "lifeless and stiff", although most modern architecture is designed in this way. Composition B comprised the same cubes, not screen-like this time, but slightly staggered. 83% now said "slightly mobile, but without direction" and 8.2% "I see movement, but cannot decide in which direction" – so 89% saw an indistinct movement.

Composition D was drawn in such a way that two groups of cubes were clearly delineated. One was a flat group of relatively loose little cubes, the other, in great contrast, a group of four much larger interpenetrating cubes. 68.3% selected "burdensome and oppressive", 24% a "sinking movement". We discovered from a further test using the same figures but with eight possible answers, any or all of which could be selected for each figure, that some 75% of the students who had selected "burdensome and oppressive" also selected "sinking movement". Both answers indicate more or less the same impression. At least 80% of the students felt that the group of large cubes was a burden to the little ones. This, together with the unevenly curved echinus moulding, is the main reason why we see the small Doric capital between the large horizontal architrave and the large vertical pillars as an antagonistic crushed zone (257).

Composition C consisted of small cubes at the bottom, changing gradually to large ones at the top. The change was so gradual that there was no grouping. In a total of four tests (we repeated nearly all the tests) between 82% and 92% selected a "rising" or "falling" but in any case a "connected movement". Only 1.5% selected "burdensome and oppressive", less than 1% "lifeless and stiff". We repeated this group again on 14 April 2000, using a regular pattern (Questionnaire XVIII). The results were similar. When the cubes lay horizontally (Questionnaire XIX) the result was again about the same. Between 80% and 90% saw movement, most of them from small to large. In all these kinds of tests, starting in 1994 with the vertically directed process from small to large (fig. C), then the corresponding picture of our test on 14 April 2000 (fig. B) up to the horizontal process on 28 April 2000 A and B the result was always that the majority saw movement from small to large. This also explains the upward movement of the diagonal rock image of figure 77 apart from the uneven curvature. We shall refer to the head of broccoli again here, in which the direction from small to large is mostly seen with an unevenly curved spiral (104). The rising movement of Manhattan's skyscrapers (254) towering towards the heavens can be explained by the different heights of the buildings, the development from small to large, from low to high. The skyscrapers of Frankfurt or other cities (e.g. Houston), which were built later, give rise much less to this striking impression of upward movement, because each high-rise building is of similar height and there are no intermediate sizes. It is precisely this difference between low and high with all the sizes in between that makes Manhattan such a visually lively city. This is caused by the skyscrapers, towering higher and higher, which grew in Manhattan

XVIII

Please compare the three compositions consisting of cubes A, B and C and decide which description you find the most appropriate for each composition of cubes (A, B and C). Please tick the description in the boxes under A, B and C, but just choose one description for each composition of cubes.

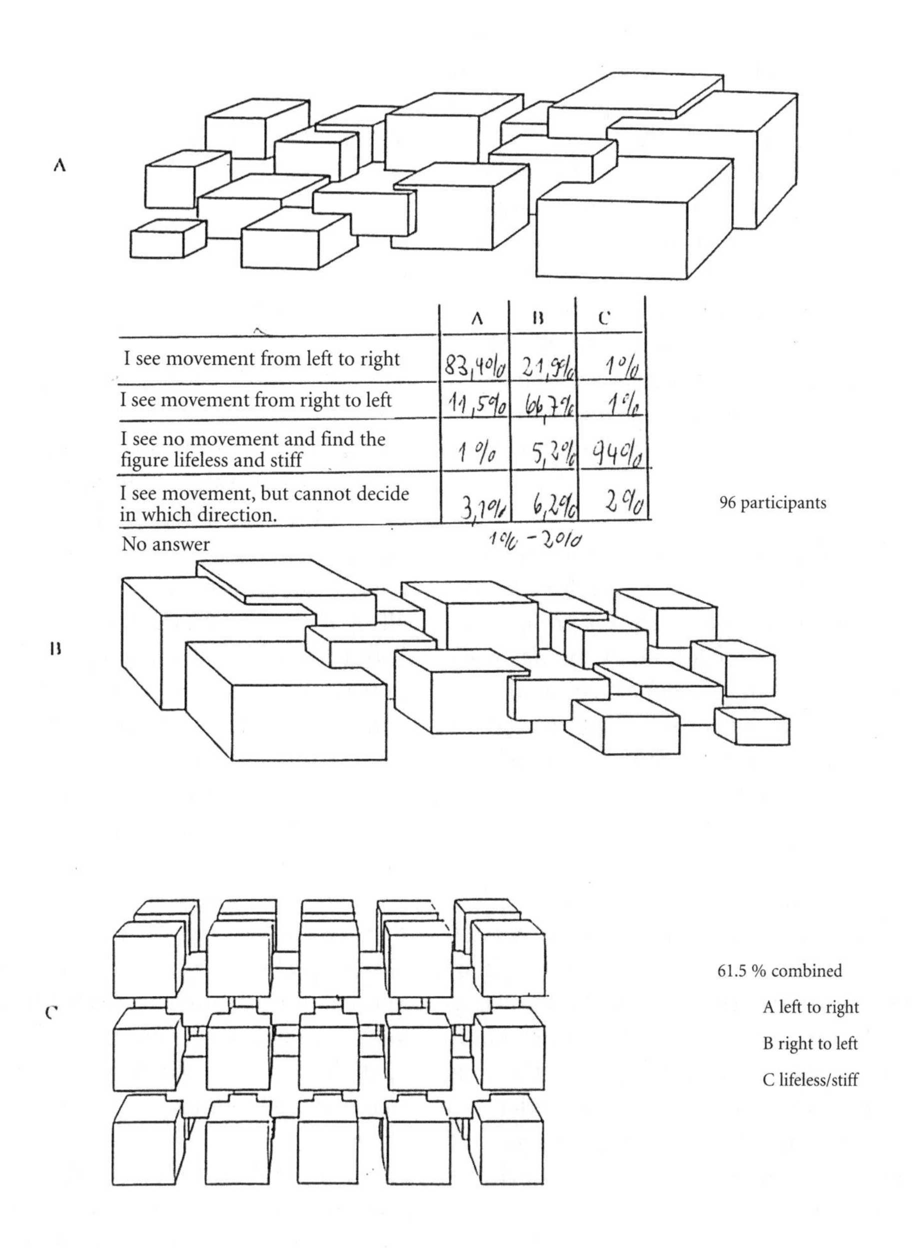

	A	B	C
I see movement from left to right	83,4%	21,9%	1%
I see movement from right to left	11,5%	66,7%	1%
I see no movement and find the figure lifeless and stiff	1 %	5,3%	94%
I see movement, but cannot decide in which direction.	3,1%	6,2%	2 %
No answer	1% - 2%		

96 participants

61.5 % combined

A left to right

B right to left

C lifeless/stiff

over a long period of time. Of course, their sheer size makes the city impressive too.

Nevertheless, in nature we do not only read directions from small to large, but also vice versa. For instance, in winter, when the trees have lost their leaves, we start with the thick trunks and proceed upward to the thinnest twigs (114, 256). When looking at the "Stadthalle" in Bielefeld, designed by von Gerkan (1980 – 1990) we also read an upward movement from large to small in the division of the windows (255). Other criteria also apply to trees, however, such as the rotation of the branches etc., just as the staggered spiral processes of Bielefeld's Stadthalle emphasize this effect. We rarely find the process from small to large and vice versa in as pure a form as in our cube compositions. Examples are: the rising smoke of a fire, developing from small to large or the feathers on a bird's wing, starting with small feathers at the shoulder and finishing with the primaries. However, the results of our tests, which we frequently repeated and varied, show that the direction from small to large dominates, perhaps because growing involves getting larger.

This brings us to a law of perception: *we comprehend as antagonistic – for instance, as bearing and burdening or being squashed (257) – groups constructed from contrasting figures – the sudden collision of big differences in size; differences in density; differences in direction (vertical, horizontal). We see gradual changes – from loose to dense or small to large – as connected movements going in the same direction.*

So, on the one hand we comprehend the burdensome effect of the horizontal architrave, the triglyph and the tympanum on the upright pillars (257), because of the contrasting directions, as bearing and burdening, but on the other hand, because of the great contrast between the long transverse architrave and the mighty pillars set against the little capitals as intervening squashed zones, and also the uneven curve of the echinus under the abacus.

We had also arrived at this result after our test on 4 November 1994 (Questionnaire V) with the altered radial figures. Alterations from dense to loose and large to small which proceeded gradually, like sinus curves, were seen as "rotating connectedly" (rotating slowly, rotating faster), while in figure A in this questionnaire, in which short, dense radials collided with long, widely spaced ones, they were seen by 78% as "rotating and suddenly stopping", i.e. directed against each other.

We have now used "dense" and "loose" and "large" and "small" to distort two and three-dimensional screens and, by contrasting the results with the undistorted geometrical screen – an image of our visual cortex – we can, we assume, recognize process, movement and spatiality. Here, too, we find that we experience undistorted geometry as lifeless and stiff. It tells us nothing, except that it exists, it has no expression. Our contemporary cities are often so sterile, because the metamorphoses of geometry are largely missing.

Chapter 12
Rotated Surfaces

There is a third way of distorting the screen, which leads to other conclusions. If we consider flat, screen-like, identical cubes, set next to each other, as in a brick wall, for instance, we then have a unit of the lowest degree. The cubes are all interchangeable; it makes no difference to the appearance. Looked at from the front, it does not seem at all spatial. There is no perceptual information about the material make-up of the screen. If we now rotate these cubes on their vertical or horizontal axes (258 – 260) we see a dramatically spiralling process, demonstrating a spatial movement of rotating cubes or surfaces. It instantly becomes a unit of a higher degree. None of these gradually rotated cubes could be interchanged without confusing the picture. Although neither density nor size has been altered in these screens, we comprehend the rotation of the cubes on their vertical or horizontal axis as the result of a process which considerably increases the impression of spatiality and, in addition, provides a key to our perceptual judgement of nature.

Every natural group of people consists of such a combination of rotated surfaces. People turn towards or away from each other, stand in three-quarter profile or profile, thus establishing communication or the lack of it. All this goes to make up a lively unit (262). A complete contrast to military formations, in which all the main surfaces are organized to be parallel and one behind the other, like the cubes in figure (249). In the third millennium, when the Egyptians for the first time in the world had their troops march in close military formation (265), they were certainly not only militarily successful because hundreds of soldiers were functioning as one man; in their lifeless, stiff, screen-like geometrical formations they must have made an additionally terrifying impression on the enemy's lively, apparently unorganized hordes. Their psychological superiority was probably dependent on this depersonalizing, geometrical appearance. In nature we are constantly experiencing this transformation of the horde or herd from rotated surfaces and dense and loose organization to parallel surfaces and identical spacing in herds of every kind. As long as they are grazing, the spaces between the animals are varied and they are a combination of rotated surfaces. If, for instance, such a herd is put to flight by a predator, the animals' flanks will all be parallel and the fleeing group will form a screen-like block, which dissolves back into varying spaces and rotated surfaces at the end of their flight. This transformation from screen to loose group with different spaces, sizes and rotated surfaces and then back again to screen, demonstrates an existential process.

At first sight the leaves of trees and bushes appear to grow opposingly and parallel. However, in fact they turn their main surfaces towards and away from each other, and that is what lets us know that they are growing and spatial (261).

Comparing the two marble statues from Athens and Samos, 560 BC and 530 BC, we can perceive how important the change of rotated surfaces is. While the Samos figure's overdress and chiton folds are indicated only by negative grooves (264) so that the folds' main surfaces, all parallel, combine to form a single vaulted surface, in the case of the girl from the Acropolis Museum in a chiton and overdress (263) the folds, roughly in the centre of her body, are directed towards the observer. They then are laid to either side, but turn in the opposite direction on her right-hand side and make another change of direction under her arms. This gentle, hardly noticeable change in the rotated main surfaces of the vertical folds lets us experience the robe's material, spatiality and life.

While in the past architects usually created plastic-spatial units with projecting or recessed rosette-like vaultings and rotated surfaces, ideal for placing sculpture to stress form and content, nowadays combinations of sculpture and architecture have become a problem, because the often box-like forms of contemporary architecture have neither spatial nor plastic qualities. I was faced with this problem when commissioned to design for the Kennedy Center in Washington (the municipal cultural centre) two architraves with figures, to go above the two main entrances (266). That is why, in designing these two architraves (each 13 metres long and two to three metres high) to go above the main entrances to the cultural centre's main hall, I based the composition of the figures on a combination of rotated surfaces. Three drums of surfaces, standing vertically to the observer and each rotating in varying rhythms in the main surface, created a strongly mobile and spatial base. Rolling over this and in its depths we see a rising and falling wave of human bodies. The bodies and the flames are made, on the one hand, of unevenly curved vaultings but, on the other hand, of rotated surfaces; a spatially structured mobile architectural element that could have formed a contrast to the flat, box-like architecture. (However, as a result of severe disagreements be-tween the architect and myself, the two reliefs were placed on plinths in the plaza opposite the halls. Here, too, they owe their presence to the spatial effect of the rotated surfaces.)

So we find that the screen is one of the most important perceptual images, most capable of variation, in the decoding of the environment. Thus it is no surprise that it is the first figure recognized by babies and that it turns up in all sorts of variations in all infants' drawings, and that the screen is one of the most important forms of decoration in early ornamentation. Even the earliest vessels, such as the prehistoric Egyptian vessel (214) in the Pelizaeus-Museum in Hildesheim and the vessel (216), 5050 BC, from Syria, bear the unchanged diagonal screen as decoration. The prehistoric Egyptian vessel, c. 4200 BC (214), combines the screen and the rectangle. The Egyptian vessel is directed upward, the Syrian, downward with its main curvature.

Evolution probably developed it as an image at a very early stage. After all, insects, such as wasps and bees, build screen-like containers with regular patterns for their larvae and food stores. The bees' information dance described by Karl von Frisch[37] also follows geometrical figures. Perhaps the very first beginnings of the declaration or reproduction memory lie here. The rosette also seems to be of visual importance for insects (in the form of

XX

Please compare both depictions of spirals A and B, read through all the possible answers and then tick just one answer for each drawing.

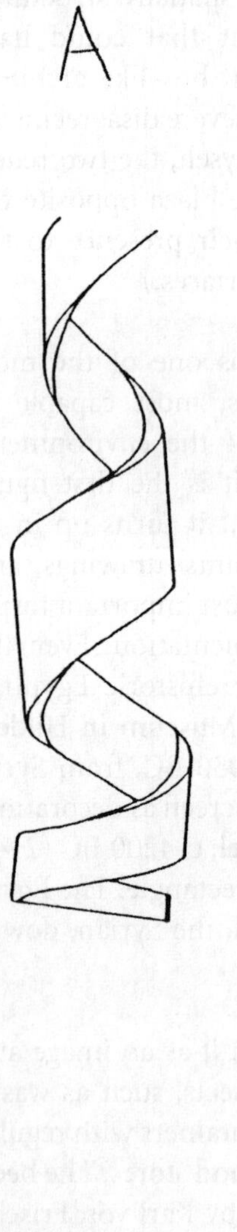

A

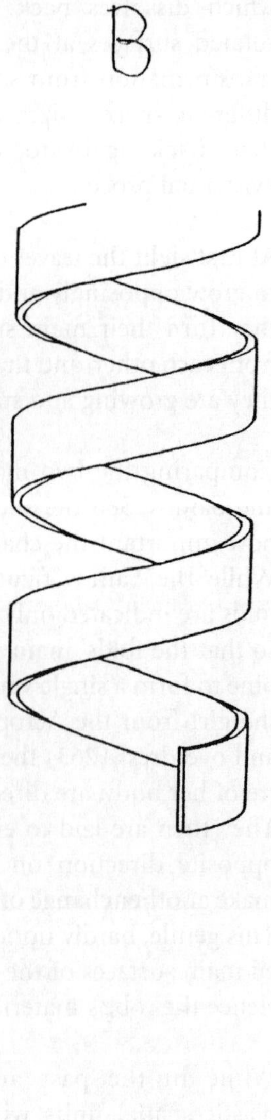

B

	A		B	
2,1%	2	rigid	61	63,6%
26,0%	25	mobile	29	30,2%
71,9%	69	springing up and down	3	3,1%
	0	cannot decide	3	3,1%
	96	Participants	96	

flowers) and for many courting birds, the rosette-like spread of the tail.

This is not such a strange idea as it might appear at first sight. After all, we know that evolution has retained tried and tested systems for hundreds of millions of years right up to the present day. It has been proved that the basic principle of the transmitting processes in the synapses of nerve cells and also several of the substances have stayed almost the same since there were such animals as the aplysia or the fruit fly drosophila. Basic chemical inventions for the short and long-term memories show little alteration, right up to the highest mammals. Why should not the same thing apply to the screen and other "orbits"? Even an anaesthetized ape still recognizes a radially organized chessboard pattern (54).

If we then include circles, spheres, spirals and circularly organized radials with even spacing as conceptual concepts, we can, by comparing this little arsenal of geometrical forms and its countless metamorphoses, with the complex forms around us, really decode an amazing amount.

That is why, at the conclusion of this train of thought, I return to the comparison of a geometrical spiral with a distorted one. This demonstrat-es uneven as opposed to even curvature; uneven, to even spacing; different, to identical sizes; and radially, to parallel organized main lines.

Four times between 1993 and 2000 we set before our students a geometrical, three-dimensional spiral with even curvature, even spacing, even coils and parallel processes and asked them to compare it with a distorted three-dimensional spiral with uneven curvature, gradually diminishing spacing, gradually diminishing coils and diverging, not parallel processes. Between 80% and 98% found the distorted spiral "mobile, springing up and down", while only between 5.8% and 11% thought it was "rigid". Between 51.3% and 74.5% found the geometrical three-dimensional spiral "rigid" and between 3% and 33%, "mobile, springing up and down" (Questionnaire XX).

Of course, in this case judgement is affected not only by form, but also by the individual's experience with the object. A person interested in technology will, of course, know that the kind of geometrical spiral illustrated in figure B can be a steel shock-absorber in a vehicle, or a spiral in a mobile reading-lamp or a shock-absorber on a pram, and as such truly mobile. Quite obviously the students had no such associations in the case of the distorted spiral, since there was a murmur of amazement in the room when I showed the ringlet (267) from which we copied the distorted spiral. Despite this, of 137 psychology students tested in 1993, almost four times as many found the distorted spiral "mobile" rather than the geometrical one, while, in the less divergent result of 2000, of 121 architecture students more than double the number saw the distorted spiral rather than the geometrical one as "mobile", even though there can hardly be anyone who has not experienced a mobile geometrical steel spiral. There can therefore be no doubt that divergence from even curvature, even spacing, identical size and the variously diverging radials compared to evenly spaced radials are comprehended as distorting movement, going against all experience with the geometric spiral. The movement of the discus thrower is also based on the distorted spiral (269). We arrive at the same result if we compare a geometrical two-dimensional spiral (271) with a distorted one (270, 272 – 275). That is why this is one of the first steps in the direction of a freer, more mobile ornamentation.

Although I have not grasped how Hoffman and others can calculate that these "orbits" are cerebral images, there is no doubt that we decode basic natural phenomena by comparing them with such "orbits" and some other Euclidean-geometrical forms. That is why psychologists – including Hoffman and Dodwell themselves – have frequently misunderstood the Lie "orbits" as one of the bases of perception. This is because they have not realized that we only compare the complex natural forms with these geometrical forms in order to reach our perceptual judgement on the basis of the differences between them; not because of the correspondence between the geometrical forms and nature. Geometry is a brilliant invention of the brain which has proved to be a suitable key to unlock the natural processes essential to our survival. There may well be similar simple basic forms in the audio area, to help us measure acoustic impressions. Obviously we can compare

[37] K. von Frisch "Die Tanzsprache und Orientierung der Bienen". Springer, Berlin Heidelberg New York, 1965 (The Dance Language and Orientation of Bees, Cambridge Mass., 1967)

parallels with identical spacing to the ticking of a clock at identical time intervals, measuring rhythmical alterations, their equivalent in the acoustic area. But I could also imagine simple acoustic basic forms, against which sounds and noises are measured. We may have here a basic principle of visual and acoustic perception.

But it is for musicians and perceptual psychologists who deal with audio impressions to decide whether such basic acoustic forms exist.

The bases for the judgement of our eyes or for the visual areas of our brains are the simple geometric forms such as the circle, square, rectangle and triangle and above all the screen. It is the difference to them that we use to measure the visual expression of the things in our environment. These Euclidean forms known as "perceptual concepts" are units of the lowest grade for our perception, because the elements of which they consist, for example the module of a regular pattern or the parts of a circle´s circumference, are all interchangeable. Changing their position does not change the whole thing. The order of each individual piece is not fixed. We have already discussed this in detail. But for Hawking, a screen is an extremely "ordered state"[38].

For our way of thinking, for our problem, however, the regular pattern attains a higher order, if it is distorted by "contraction and expansion", "enlargement and reduction" or "rotated surfaces", if the individual elements are interchangeable no more or only to a low degree. The sphere, for instance, that looks the same whichever way it is turned, that is not fixed to a point in space, becomes a unit of a higher grade through the directed expansion of its volume and distortion to a drop. It is possible for these types of distorted perceptual concepts or geometric Euclidean forms to combine visually with each other and thus, for perceptual thought, their grade increases to a high point. From this point the curve falls off again to a completely amorphous state (e.g. figure 62, there are certainly even more amorphous states), so that the visual unit again goes back to its lowest point. This curve of units of lower or higher grades also describes the curve of visual information for our perception. This organization system differs fundamentally from the system of entropy that is one of the bases for Hawking's theory of the expanding universe. There are obviously very different points of view regarding the formation of organization systems for spatial and chronological phenomena. Maybe they could even stimulate each other, it seems to me that they could. The organization system we have shown here makes it possible for us to interpret all the phenomena in our environment on the basis of their difference to the Euclidean forms, the perceptual concepts, or for us to understand information visually.

These geometrical conceptual concepts and their metamorphoses are also an important formal prerequisite of understanding works of art and craft. Nowadays the formal contemplation of art is all too often dominated by aspects related to its contents and history.[39]

[38] Stephen Hawking "The Illustrated A Brief History of Time". New York 1996

[39] Tonio Hölscher "Klassische Archäologie - Grundwissen" Darmstadt 2002.

PART II

Chapter 13
Form and Movement

The first thing a baby sees is movement. Initially the baby's movements, known as mass movements, are very uncoordinated. In its second year, its movements are more directed, as it ventures further into the space around it, and finally starts to scribble (16 and 17). In these scribbles we can clearly trace the infant's moving hand or circling arm.

For perception there are basically two kinds of movement. So far we have not differentiated between them. There is an object's actual movement in time and space, without changing its form, from place to place. A good example of this is the motor car. The groups of neurons or visual cortical areas that register such movements have been identified fairly accurately. Then there is the kind of movement that distorts form, for instance a fruit growing or an animal running. These latter movements frequently leave permanent alterations of form behind, which inform us about past movement. We see the infant's hand in action in the ups and downs, backward and forward strokes and the tangle of lines. Basically it does not matter whether we actually watch them being produced or just see the result.

Apart from these, there are real movements, alterations in time and space, which happen so slowly that our eyes cannot see them: for instance, the forward movement of the hour hand on a clock or the growth of a plant. But we can see from the shape of fruit (63, 74, 75) – we need not even compare younger and older stages – that the fruit has grown and will possibly continue to do so.

There are very simple processes: slowly falling snowflakes in a calm, drifting clouds, moving traffic and many other things that we can also imagine very accurately. But there are also incredibly complicated movements, which we cannot properly register even if we watch them, and which we can only describe vaguely afterwards. Think, for instance, of the completely different artistic depiction of equine movement before and after the invention of stills. A horse's stance in equestrian statues of the Renaissance, galloping hordes of horses and riders in European or Chinese pictures, differ from a horse's natural movements as we have come to know them from stills (276, 277). Horses painted by Toulouse-Lautrec (281) or Liebermann after the invention of photography look different from those painted by Benozzo Gozzoli, Velazquez or the Chinese (278, 279, 280).

Almost all the movements of the higher mammals, from predators via apes to humans, are extremely complicated, difficult to describe and so cannot be reproduced by the declaration or the reproduction memory. All we find there are very simplified movement schemata. There will be more on this later.

Let us start with the simplest movements. These are the processes which make full use of the available space, e.g. falling snow. If we shut our eyes, we can watch such movements almost as if we were looking at a film: drifting clouds; smoke rising from a fire; cars driving fast or slowly, or in an uninterrupted line; a revolving Ferris wheel at a fairground. Rising or falling movement which takes up all the available space, horizontal displacement, rotation, expansion and contraction are the simplest processes. We can conjure them up before our inner eye and they are symbolized by the "orbits" of Lie mathematics that have already been mentioned.

It is because they are so simple that they are the processes depicted by infants beginning to scribble. Backwards and forwards, hither and thither, circular movements and finally lines which are screen-like and which cross each other at right angles – the infant depicts the basic processes, even if, while doing so, it talks about other, much more complicated ones. I pointed out the connection between infants' scribbles and the basic processes in my book "Gestalt, Bewegung, Farbe" first edition 1975, third edition 1984 and in 1983 John Matthews also very precisely described children's movements before the scribbling stage as the existential conquering of space[40]. Over a long period he carefully observed what his three children did before they started to draw on paper. He described three basic movements which laid the foundations for the children's drawing activities. The first is an up-and-down movement, a thrusting movement of the arm and hand, which collides with the object, the floor, the wall or later the paper almost at right angles. He called this movement the verti-

[40] Claire Golomb "The Child's Creation of a Pictorial World". Berkeley, 1991

cal curve (21). The second is a wiping, stroking movement directly above the horizontal surface, which he calls the horizontal curve (17). In this way the child comes to know the space around its body. He calls the third movement a push-pull action (16). The infant grips a pencil or some other marker – it may well be a spoon – pushes it to arm's length, then changes direction back to the starting point. This action creates lines running to and from the body. Push-pull actions contain a series of forward and backward movements. The infant's initial push-pull movements leave no visible trace, but the pencil leaves a clear visual mark (16). These three types of body action, carried out with a tool, a brush, or a spoon in porridge leave, to quote Matthews, "a trace of their passage through space and time".

This brings us to our subject matter. Infants' first scribbles are not some sort of hand movement exercises – apes with pencils draw very similar figures, and with their nimbleness would hardly need such exercise – they are the expression of spatial conceptions, of what I called at the beginning "basic processes", to which all movements refer and are traced back, even the most complicated ones made by the higher mammals.

From the beginning of their second year, infants work with brush, pencil and colour and discover the effects of their vertical strokes and horizontal sweeping movements (21, 17), which create various pictures.

At the very beginning of this book I described how, from the basic processes, the basic forms – circle with its radials, screen and rectangle – arise in individual steps which can be minutely observed. The tadpole emerges from the circle with external radials when, at right angles to each other, two of these are lengthened to form arms and two, legs. Then, finally, these latter are divided by horizontal lines into rectangles representing neck, body and legs (36). The central symmetrical figure is thus combined with the ladder or grid system. The child now employs the original spool (18) as hair for its female schema (36). The ladder or screen system is used to represent a spruce (43) or a high-rise building (39). The world around the child is newly reconstructed by means of the basic geometrical forms it has invented and which it then deposits in its declaration or reproduction memory. This memory content needs no models and can be reproduced at any time. This phase usually lasts till the tenth or twelfth year. This is how the symmetrical nude schema comes about (42), as do the schemata for clothed people (45). For many of the objects surrounding it the child now develops simple form schemata, combining Euclidean geometrical forms, and not expressing any kind of movement. A tree schema can just as well be made using the rectangular ladder schema (358) as the circle and rectangle (357, 360).

As I mentioned in connection with the drawings of the spruce, we asked a group of children between their sixth and thirteenth year to draw elephants. The pictures all resemble each other in that the elephant's body is roughly a rectangle, sometimes also developed from the segment of a circle, the jointless legs are fixed on at right angles, occasionally the head lacks its own geometrical form (288 Olga, 13 years old) and is only drawn onto the body with an eye-dot and ear, or else the head is put in front of the rectangular body as an extra, geometrically much simplified form (287 Natalie, 12 years old, 282 Joachim, 6 years old, 286 Rasa, 11 years old, 289) etc. Of course the Euclidean geometrical reproduction of natural forms is narrowly limited. Geometrical schemata frequently resemble each other; as they get older children become more and more aware of these forms' difference from the intended natural model. The result is that at some point in their twelfth year they usually lose interest in drawing and that is the end of the development of the visual reproduction memory. This is when school "art education" ought to take over.

In November 1999 we asked 145 students to draw an elephant or a cow from memory; always one drew an elephant and the neighbour a cow, so that everyone had to depend on his or her own memory and could not copy. 87% of the 86 students who drew an elephant used a purely geometrical schema (290 to 294) – completely immobile – some even used a single form for head and body, drew in eyes and ears and added a trunk in front. Just like the six to thirteen-year-old children, the body was usually geometrically rectangular or developed from spherical or ellipse-like forms, supported by rectangular legs stuck on at right angles. 4.4%, i.e. four students, managed rather more natural depictions, i.e. slightly distorted basic geometrical forms (see 295, 296). Six students, or 7%, produced cartoon figures, which they must have practised (297), but were purely geometrical.

Only 59 students drew a cow. Apparently many found elephants easier; the trunk makes them instantly identifiable. Of these 59 students, 51 constructed their cows from the simplest, slightly curved or simple rectangular geometrical forms (300, 303, 306), lacking almost any representation of mobility; 28, i.e. 47.5%, depicted their cows without horns. This means that nearly half the students had stored hornless cows in their optical memory – they thought only bulls had horns. That really cannot have anything to do with drawing ability (300, 302, 304, 306). 10.8%, i.e. six students, produced fairly characteristic distortions of the geometrical forms, even putting joints in the legs, thus showing that they had somewhat differentiated memories (298, 301). Two, i.e. 3.4%, had also practised cartoon figures (305). Some students bent the cow's forelegs the wrong way (304, 306); again, hardly a question of drawing ability but more of a faulty visual memory.

Let me point out once again that these young people (all over 20) were students of architecture, a faculty in which some 1,500 students – some 165 in the first year – are always enrolled. In the last three years only six, i.e. 1.73%, of 436 exam candidates failed to get their diploma. Is the reason for the wide-spread monotony of modern architecture to be sought rather in this faulty training of their visual abilities than in "functional/constructive" building styles? A person who uses the same geometrical schema to draw a spruce, a fern-leaf or a palm will also develop a town hall, a house or an office on the same schema.

As a rule our intellectual judgement – if we are not dealing with very abstract mathematical-physical matters – consists of consecutive, constantly alternating, perceptual and logical partial judgements, which lead us to our final judgement. Rational judgement is extremely dependent on memory content. If our perceptual reproduction memory atrophies, it damages our general ability to form judgements. We can grasp neither the whole nor the detail if the content of our reproduction memory is limited to such simple geometrical forms as those in the children's and students' drawings. The schemata stored in our memory all look alike. Memories like these make it impossible to form a differentiated perceptual judgement. Apart from the question of whether we are at all interested in art – a subject which schools perceptual thought, the ability to memorize pictures – our educational politicians, who usually only pretend an interest in the arts for the sake of the media, fail to appreciate the importance of perceptual thought for the entire thought process. If it atrophies our entire thinking is endangered.

Even if the depictions of animals on the Korean vessel from the three-empires period, early fifth century AD (308), so far as the characteristic differences between stag and horse are concerned, are rather better presented than in the drawings by the children and the students and also show more leg movement, it is still true to say that these drawings from the early stages of Korean culture are constructed from simple geometrical forms and differ only slightly from each other. Here, too, the neck of the vessel and the animals' bodies are decorated with screen patterns. The same applies to the Egyptian vessel with hunting scenes, middle of the fourth millennium BC (326), even if the form of the Egyptian vessel, four thousand years older than the Korean, is far more perfect. We can dismiss almost completely any idea of a connection between the cultures. The stag on the ritual Korean vessel, fourth to third century BC, (310) shares the level of the drawings by the children and the students. It is rectangularly constructed of the simplest geometrical forms. The earliest Chinese drawings depicting animals (fifth millennium BC) are almost purely geometrical, so that the frog with two legs could just as easily be a beetle (327). The hippopotami on the predynastic Egyptian dish, 4000 to 3600 BC, consist simply of two slightly curved rectangles between two blossom or star-shaped rosettes made of three or five parallel lines (328). The Chinese elephant, 1000 BC, (329) decorated with circles and "orbits", resembles in principle the students' drawings (294). This also applies to the rubbing of a Chinese elephant relief, thirteenth to eleventh century BC (331). Please note the rectangular spirals organized to resemble a screen.

The same also applies to the animal depictions from Moravia, Hallstatt era, 700 BC, (323, 324) and to the Cypriot horse, c. 700 BC, carrying amphoræ on its back (322). Again and again in these depictions of animals we come across the already mentioned Lie "orbits": here, parallel lines and the eye as a double circle. This also applies to the horse on the Etruscan vessel, c. 100 BC, (321). There is no end to the possible examples from the beginning of every culture on every continent.

Also on the Greek amphoræ, 1050 to 700 BC, in the geometric style, we find such depictions of animals and people; the horse's body almost rectangular, decorated with a chessboard pattern (318). The horse from Trondholm, Denmark, part of a chariot of the sun, has been somewhat distorted – is rather more mobile – but is still basically geometric (313). Also slightly distorted, but still primarily geometric, the bronze statuette of a horse from Sparta or Tegea, 725 BC, late geometric (320). The goat on the Cypriot vessel, seventh century BC, (319) also belongs in this category. He not only has the Lie ornaments according to Hoffman on his rectangular body, but also on his strongly distorted wings. The vessel is divided up with parallel lines. The Egyptian hippopotamus, c. 3000 BC, (307) was also developed from cubes, slightly distorted by the basic processes. Its legs are put on at right angles. The Egyptian hippopotamus from a later period, 1850 to 1700 BC, Middle Kingdom, is more definitely constructed from distorted geometry and thus more mobile, but also decorated with various sized circular "orbits". The triangular depictions of the lotus flowers are also basically sections from such circular "orbits" (325).

The pre-Columbian dish from Panama (330) is decorated with a serpent-demon consisting of a circle with external radials. All over the world the initial depictions of animals were geometric. For instance, the elegant, fanciful African antelope-being, also decorated with a triangular screen pattern, has been almost entirely reduced to a purely geometric body. Obviously these and similar sculptures, even in their smallest details, influenced Picasso and others (312). In Africa the geometric schemata, admittedly distorted by the simplest basic processes – for people, too – dominated art into the first half of the twentieth century, and so had an enormous influence on cubism, from Picasso to Matisse, Zadkine, Lipchitz and Laurens. The protogeometric tenth-century BC Athenian stag, – completely immobile – belongs to this category, as do its decorations (317). The tenth-century AD Peruvian jaguar is constructed almost entirely of rectangles, squares and triangles (333).

After the Second World War, when most art journalists wanted abstract art to prevail – frequently they were the same people who, under the Nazis, had defamed the "degenerate" artists and supported Germanic art – we often heard the argument that the geometric abstract style was not the beginning of art, but its climax. The beginning had been the naturalistic art dating from between 15000 and 12000 BC discovered in Altamira and other caves. Even if I ignore the fact that these pictures are not naturalistic but close to nature, and also, unlike nature itself, created from geometrically distorted forms, what we are being told here is a gross distortion of history. This cave art of the higher hunting culture between 15000 and 12000 BC in Altamira and other caves in north-west Spain and southern France was in fact the end of a stone-age artistic development, which had begun c. 75000 BC, probably just like later cultures with zigzag lines (archaeologists call them macaroni lines), ladder and comb shapes, (315) lines crossing each other at right angles, that is the screen, and geometrically only slightly distorted animal depictions, which nearly always had only two legs instead of four (311). But as this beginning of the late Magdalenian age was almost as far removed in time as the Magdalenian is from us today, not much of its art survives. Some of the caves have collapsed, the early drawings are covered up by thick calcareous deposits or later artists simply painted over them. The beginnings are so nebulous that we shall not be including a consideration of early Palaeolithic art. I have only mentioned it because this argument, which is completely topsy-turvy, is always cropping up. The pictures from the Magdalenian age, fifteenth to twelfth millennium BC, are not a beginning, but the initial end of a cultural epoch. In the history of mankind there have been at least three or four such epochal cultural interruptions. Although this is now also put into doubt by the newly excavated city in Eastern Anatolia – even though it only consists of a collection of temples (c. 9000 BC).

The depiction of the human figure also starts everywhere with almost or completely immobile purely geometric schemata. Particularly cubic in America, an Ecuadorian stone idol, c. 2500 BC, which then, step by step, still very cubic but more differentiated, was transformed into the female stone idol from Mexico, 800 BC (334, 335). In many places in the world this stage starts around 2000 BC, for instance this stone idol (336) from south-west Spain, developed not from a cube but a geometric cylinder. Note the eyes: circles with radials. This is a clear example of the fact that initially artists were interested not in depicting visual impressions but in reproducing our brains' geometric perceptual images. People certainly have eyes, but these do not look like wheels, however abstract the depiction. We shall come across other examples of such symbols combining circles and radials, each having a different meaning – up to the oeil-de-boeuf windows of the patient with Alzheimer's disease. The arms – it is impossible to say where arm, hand and finger start – consist, like the hair, of parallel lines. The menhirs from France come from the same time (337). Like some of the elephant drawings by the children and students the head is not depicted as a separate unit, but as a slightly pointed rectangle, eyes and breasts as drilled positive or negative little circles and the necklace a multiple circular "orbit". The arms, bent at right angles, finish in parallel fingers which are all the same length, as are the toes at the end of the rectangular legs. The human figure was constructed using the simplest two-dimensional geometric forms. This also applies to the cruciform female idol from Sardinia, c. 2000 BC, which, however, consists of a cylinder connected to an almost rectangular slab, with breasts and nose attached (338).

In the eastern Mediterranean area this development started with the violin idols, where, as early as the fourth millennium BC, at least an attempt was made to show the shape of the breasts and hips, even though there was no head. There were also purely rectangular, and thus expressionless ones (cf. 339, 344). The rectangular bearded Egyptian figures, first half of the fourth millennium BC, almost without details, are examples of the purely geometric, scarcely divided beginnings of figural depiction (made of hippopotamus tusk and slate) (353). Something similar can be said of the cylindrical, slightly unevenly curved female Egyptian sandstone idols from the third millennium BC (354, 355). In comparison, the female idols from Moravia dating from around 2000 BC are astonishingly vivid, although their arms stick out a right angles (343).

Fairly soon, shortly after 4000, the violin idols of the eastern Mediterranean, where Greece is today, developed into a form clearly recognizable as a female figure, with a head, breasts and thighs, distorted by directed basic processes. Around 3000 more and more details appeared, together with more strongly directed curved, and thus expansive vaultings (340, 341). So it was in the area of Greece that we find the earliest distortions of simple geometry to unevenly curved vaulting – certainly not later than in the Egyptian-Nubian or the Chinese area. Let me draw your attention to the Egyptian porphyry statue (345). Here, too the sculptor started with a cube, though with a clearly separate head, and began to distort the purely cubic form into an upward directed curve with corresponding – naturally symmetrical – details. The clay statuette (prehistoric Greece) of a woman sitting on a chair nursing a baby (346), also decorated with parallel lines, represents a similar beginning of the metamorphosis of Euclidean geometry into an organic form, at least into directed vaulting.

Although differently constructed the clay statuette of a seated man from Nigeria, 200 BC, (347) demonstrates a similar stage. The body and the neck are purely geometric cylinders, The head, however, is a slightly upward unevenly curved and thereby directed form. The arms are almost rectangular, but this schema is given a certain human mobility and expression by the legs bent at an acute angle and the modelling of the facial features.

But before taking the step from the purely, or almost purely geometric form schema to the distorted organic figure, we should stop for a moment and look back. These early historic ornamental forms, sculptures of animals and people were not created by children who had first to conquer the three-dimensional aspect of their environment and who began by reproducing the geometric images they had developed, but by adults. Even the scribbled lines (macaroni) on the mud walls of the early ice-age caves were presumably drawn by adults and not children.

In my opinion reproducing mankind initially reproduces the geometric forms developed by his brain and from them constructs his environment anew in a very general and universal way. Initially these visual images of our brain seem to be more dominant and simpler for us than external impressions. But they set narrow limits to reproduction and interpretation of the world. We have called upon many examples to show how small the store of Euclidean forms can be which we use for interpreting the world and comparing with natural forms, especially when we multiply these geometric forms by combining them to make new figure schemata. But on the other hand the possibilities for using them exclusively to reconstruct the world are severely limited by this small number of basic geometric forms. There are peoples who were happy to do so for thousands if not tens of thousands of years and others, more interested in an individual reproduction of the environment, who therefore distorted their geometry faster and more drastically in order to create in their art, too, forms closer to nature and spatially more mobile. Besides interpreting the expression of natural living things (spruce, palm, willow etc.) we keep returning to art as well, because – especially in its beginnings – it allows us to draw conclusions about our visual memory and thus about perception and consciousness, too. We must not imagine the pre-archaic sculptor or painter working from a live model. He presented his memory picture of the world (Greek eidolon = picture, form) and his work was accepted, because it corresponded with pictures in other people's visual declaration memory.

But no matter whether we are looking at nature or art, we read the processes that have distorted them in their differences from the simpler forms, the metamorphoses of geometry. This tells us their fate, their real or imagined past, present and future. We do not need an interpreter to understand their information. Undistorted geometry, an invention of our brain, contains little or no information about life, because it is by comparison with it that we recognize and judge the vital processes of the various objects around us. It is, for our visual recognition, the standard by which we measure everything and that is why it says as little about itself as a yardstick does about a person.

That is also why the very early native-American idols, rather like gravestones in their undistorted forms, are far removed from life and almost expressionless. The rigid, symmetrical eyes may have a certain magical effect, but I rather think there is a fashion to overvalue such things. Such geometric-symmetrical images are more expressionless and lifeless than magic and terrifying. Even though at the end of the Stone Age the metamorphoses of geometry, and with it the expression of vital processes in art, began in the Greek-Cypriot, Moravian, Egyptian-Nubian, Mesopotamian and Chinese areas, purely geometric depiction, on the other hand, with its weakness of expression, still exists today, either because of primitiveness or for cultic reasons. This applies to the grave guardians, c. 1000 AD, from Mexico (348), to the geometric x-ray depictions of men and animals on a not very old native-American bag from North America (351), to large parts of Australian art and to many geometric depictions of people and animals by children, and also by most present-day adults.

In contrast to this, many African sculptures are constructed of geometric or similar form schema, but have more expression because of the contraction and expansion and enlargement and reduction of their separate elements (350) and of course because they are directed by vaulted distortions of geometric forms. That is why, at the end of the nineteenth century, they had such a great influence on Western art, which had been increasingly losing its orientation and growing more naturalistic (up to the shaping of figures in moulds – the cast).

But we recognize the vitality of our environment, the growth, maturity and decay of plants, their breaking through the earth and fading, by comparing them to unchanged geometry or to a corresponding simpler form. There is no doubt that the basic distorting processes, expansion and contraction, horizontal displacement, rising and falling, the processes that totally fill space, are the most important and frequent happenings around us. In nature we are most frequently confronted by drifting clouds, the spreading and growth of bushes and trees, vegetation as a whole, fruit swelling, buds opening, the vital dynamic of leaves, the stretching and blooming flesh of youth and the decay and sinking of age, landslips in mountainous country. We recognize this growth and decay through the distortion of geometric forms by the basic processes. They show us life by comparing it to our brain's lifeless geometry with its identical sizes, identical spaces, identical degrees of curvature and the unaltered straight edges and right angles.

But just as most children stop drawing in their twelfth year because constructing our environment from simple geometry and geometric schemata loses its charm – they realize that things that look very different in reality all look so similar, can scarcely be told apart in their drawings – in a parallel development our perceptual judgement, our understanding of life, of growth and decay slows down, if our visual memory stores predominantly geometric forms and schemata resembling geometry.

That is why, on the basis of these forms, we develop through practice more complicated schemata with which we can compare and by which we can judge the infinity of forms around us at a higher level. There is an alarming curtailment of this ability in present-day people. I call this the visual "Kaspar Hauser effect"[41] which affects people whose artificial, self-created environment consists, from computer and television to refrigerator, car and high-rise building, more or less exclusively of cubes and grids and whose vital impressions are limited more and more to two-dimensional television and the computer. The visual "top-down" and "bottom-up" development is broken off too early, the self organization of the brain possibly stops at a halfway point[42].

In order to recapitulate let us now analyze three plants as examples of what I have so far said about the metamorphoses of geometry by means of the basic processes.

The simplest form schema is constructed of geometric parts (357, 360) – a tree made of a circle and rectangle or a rectangular screen or ladder-system (358) – as symmetrically as possible. If, for example, we look at the breadfruit tree[43] (361) we realize that it has grown very crooked and diverges widely from the schema norm, that it was probably so strongly distorted by an external force – the prevailing wind on this coast. Even while actually perceiving visually we are forming a rather complicated purely perceptual judgement.

We know from neurophysiology that memory content is stored in those parts of the visual cortex which also initially take in the corresponding visual impressions from the environment[44]. So as it is seen, the crooked tree (361) is compared with the symmetrically vertical form schema. But that is not sufficient to understand the complicated perceptual expression. If we simply draw a geometric tree schema crookedly, distributing the branches in an even radial manner (361 a), we feel no trace of the drama of distortion which took place be-tween the growth of the tree and the prevailing wind. It is just a crooked tree schema. If we then distort this depiction which, compared to the natural model, is still very schematic (361 b) so that we increase the treetop's degree of curvature against the wind direction and let it decrease with the direction of the wind, the impression we get is already much more dynamic. If we then also alter the spacing between the radial branches to be wide at the bottom and narrow at the top and if we gradually reduce the size and strength of the branches from bottom left to top right, then although we have not presented the drama in detail, we have stated the basics. We see a tree that, in growing, has struggled against its deformation by the wind (361 c). The drama between growth and wind shows the metamorphoses of the geometric tree schema into a natural tree, or the comparison of the natural tree with the geometric tree schema, i.e. the comparison[45] of uneven with even curvature, of uneven with even radial spacing and of different with identical size. This is how we arrive at our perceptual judgement, which we have already tried out on various examples of trees (spruce, palm, willow and breadfruit), a fern-leaf, fruit and a ringlet of hair and also on human faces.

Nature's most frequent phenomenon – growth, maturity and decay – in which the most important basic processes participate, is depicted anew and differently billions of times and over and over again by the leaves on trees and bushes. Let us, before going on to the next step, take another final detailed look at leaves and branches. As I write these lines, spring is beginning, it is the middle of April and everywhere the leaves are unfurling.

Let us first be clear in our minds what it is that lets us recognize growth and the spreading of plants and leaves, when looking at a group of rose bushes which started to put forth leaves about two weeks ago. The main shoots (362) grow apart radially upward from a common root, but differently spaced and in the three dimensions of their space (cf. Questionnaire V with different radial spacings). The subsidiary shoots taper off from the main shoots. Their leaves are opposite each other in pairs and each shoot ends in a single leaf, so basically the leaves are paired, and organized to be identically dense. However, the picture is much altered by the fact that the main shoots growing radially from the root diverge more and less from

[41] According to the encyclopaedia "Grosses Brockhaus": Kaspar Hauser was a foundling supposedly born on 30 April 1812. He spent the first sixteen years of his life in a dark room, causing his intellectual development to remain very limited.
[42] Engel, Fries, Singer in "Nature Reviews/Neuro science", op. cit.
[43] Jürgen Weber "Gestalt, Bewegung, Farbe", op. cit.
[44] Rita Carter "Mapping the Mind". London, 1998
[45] Wolf Singer "Zur Selbstorganisation kognitiver Strukturen", in "Gehirn und Bewußtsein", op. cit.

each other and also intersect. This gives rise to contractions and expansions in the area of the main shoots, which we perceive as spatial movement. Since all these main shoots branch off into subsidiary shoots with paired leaves, expansions and contractions also come about among the leaves, even though they grow symmetrically. There are also intersections, in contrast to some which can be seen singly. Although the leaves are organized as pairs, they change direction, are not parallel and turn, some more, some less, towards the light; not to mention the fact that some leaves have already decayed (363 to 365) or have pushed against each other.

Each leaf develops from its main axis in two halves. The main axis itself is unevenly curved; it begins growing with a main curvature on a twig, later the leaves curve most strongly towards the end and thus we read the uneven curvatures as dynamic movement from the shoot outward (cf. Questionnaires I and II on unevenly curved vaultings and lines). Right at the end of some shoots at the top of the bush we see in figure 365 some leaves whose buds have just unfurled. The leaf halves pull apart from each other, but in doing so turn towards each other, so that usually the unevenly curved main axis becomes the keel of the leaf thus created. But simultaneously the halves curve rectangularly towards the midrib, first concavely inward and finally, as they continue growing, convexly outward. This double curvature of the whole leaf, which alters as it grows, not only gives it great rigidity compared to its strength but also expresses a dynamic movement from the shoot forwards and sideways, a spreading of the vegetation (371). Besides this, depending on the way the light falls or their meeting with other leaves, the individual leaves turn at different angles, so that there is an interplay of rotated surfaces between the pairs of leaves and the leaf at the tip. (Not always clearly visible in the photographs.)

Thus the directed basic process of growing and spreading is expressed and seen by us through the interplay of radial main shoots, stretching away from each other, which spread into the three spatial dimensions at different angles and in uneven densities, and the doubly unevenly curved surfaces of the leaves, together with the interplay of the rotated surfaces which also create spatiality. These three categories, applied to various types of plant, tree and leaf, are responsible for an incredibly multifarious range of expression, which we recognize by comparing them with the schematism of simplicity of identical angles and densities, even curvatures and parallel surfaces, i.e. with our visual memory's basic geometric forms.

As a further example of this multifariousness, which depends on such simple categories, let us consider an evergreen – the cherry laurel – (367) whose leaves have already completed their growth, so that the middle axes always curve towards the end, thus indicating the direction of growth (cf. the direction of unevenly curved lines, Questionnaire II) and whose two side surfaces are curved outward vertically to this midrib, also usually with the main curvature towards the edge (366). So the leaves are doubly convexly curved. However, the two halves of the leaves oppose each other at lesser angles, the double curvature is flatter than that of the rose leaves and thus has a different expression. While the branches diverge in different directions, the budding umbels grow almost vertically and parallel, but with various spacing. The main curvature of each tiny bud is upward. This indicates that they are about to open, to expand (367, 368).

Undoubtedly the basic processes of rising and falling, horizontal displacement, expansion and contraction, are the events which provide us with the greatest experience of space and give form to the entire environment. They are the most important precondition for our visual perception – as is light.

The infant's first task is, with its movements and body sense, to become a part of the basic processes. From them it later develops, as visible forms of the basic processes, Euclidean geometry – parallels, ups and downs, the right angle and the rectangle, spools, spirals and the circle with its radials, and the screen with identical sizes and spaces. The basic processes make the whole of space available. What the circle and sphere plus radials, the rectangle plus cube, screen and three-dimensional grid manage is impossible for the triangle plus cone and pyramid. This form always indicates only a segment of a space. That could be one explanation for why the triangle is the last Euclidean form the child discovers, round about its fifth year.

In the visible natural world around us, not in objects we have created, though there are occasional approximations, in a strict sense this Euclidean geometry does not exist. The celestial bodies are approximate spheres, circle, sphere, screen, cube and geometric spirals exist in the microscopic and ultra-microscopic realms, but these are discoveries made well after the animal/human visual cortex, millions of years ago, developed the geometric forms as a symbol of the basic processes. Using these simple geometric forms the infant then newly constructs in his reproduction memory our environment's far more complicated phenomena as form schemata, just as primitive people did – nor do most present-day adults proceed far beyond this stage.

On the one hand, visible environmental phenomena are in stark contrast to our brain's geometric images but can, on the other hand, be compared with them, derived from them. We recognize the effect of the basic processes in these phenomena because of the differences and through comparison of the natural forms with these geometric images. Just as the leaf curves unevenly in its two main vaultings, which develop vertically and radially to each other, we recognize by comparison to an even curvature, or to a straight line, the direction of its growth. In the same way we realize the direction of its growth in the uneven curvature of its outline, the sudden widening of the leaf from its stalk and the second main curvature at the end, usually finishing in a point at the end of the midrib. The example here is a beech leaf (372), which we compare with a distorted ellipse rather than a circle. Leaves, especially beech leaves, belong to the rare natural examples of a large number of almost identically spaced parallels in the form of veins. The lime leaf (373) with a strong, often unsymmetrical curvature of its outline as it develops from its stalk, and which often, after a flat curvature, passes over into a second main curvature in the forward-pointing tip, is more likely to be compared with an evenly curved circle. While the lilac leaf (375) starts with a strong curvature at its stalk and ends in a flat curvature at its forward-pointing tip, the magnolia leaf begins with two diverging straight lines or completely flat curves and again ends at the tip with a strong curvature (377).

Sinuate leaves such as the oak leaf (378) point with their main curvature and the individual segments, as far as the total outline is concerned, in the direction of growth, while the usually five-part vine leaf (374, 376) starts at the stalk with a strong curvature and in its total outline finishes up rather flatly at the tip. On the other hand, the individual segments are frequently most strongly curved once again in the direction of growth. In the case of many leaves this is emphasized by serrated tips pointing in the direction of growth.

Thus we see that the result of the questionnaire in which c. 30% of those asked interpreted the streamline as flowing from the strong to the flat curvature, while c. 70% said the opposite, was completely justified. If, as in the case of the lilac leaf, the strong to the flat curvature is combined with an arrow-like, forward-pointing tip, then the direction from strong to flat curvature is clear.

There is no end to these examples of leaves with different vaultings and outline curvatures. They all show that the uneven vaulting or line is interpreted in the light of the even curvature or straight line, and its dynamic is dependent on the difference. There are millions of variations, billions of which surround us.

The simple geometric forms arose out of the basic processes, just as the Lie "orbits" I have mentioned mean both the basic processes and the geometric Euclidean forms, which in their turn are distorted in such a way that we recognize growth processes in these distortions. So, on the basis of a few basic geometric forms, the invention of our visual cortex, we judge an inconceivably large number of phenomena and their modifications.

Chapter 14
The Metamorphosis of Geometry in Egyptian Art

Human and animal form schemata, idols, are given life initially by the basic processes, not by specific movements on the part of the humans and animals. If we want to know what our visual memory can do and how and what we perceive visually, if we want to see these things objectively, then we have no choice but to concern ourselves with art. This is because the products of art show us what we remember and also how reality changes in our perception. Of course, it is true that only a few people can reproduce, depict or interpret reality, but it is just as certain that until a few decades ago the non-creative could also identify with these products of art. Indeed, as a rule, unless it was pointed out to them, they saw scarcely any difference between the real world around them and its artistic re-creation. Otherwise, for instance, it would not have been regarded up to the Renaissance, up to Vasari, as the highest aim of art to imitate reality. That was the most essential criterion, even though artistic depictions, whether archaic or as close to nature as in the high Renaissance, were always basically different from the intended reality – not just materially but also because of their geometrization. The visual cortex does not reproduce like a camera but reconstructs reality from its visual images.

Let us start with the Egyptians, since theirs is one of mankind's earliest cultures and obviously developed in complete agreement between the artists and those who looked at or used their art – which continued for 3000 years without changing its identity. Some examples of their earliest art have survived, from the "Thinite" time in Egypt, 3000 to 2778 BC. I have already mentioned the porphyry statuette, 3000 BC, (345) and should now like to again mention the lion from the Berlin collection as an example (379). It needs no particular schooling to realize that this seated lion is constructed from almost rectangular cubes whose edges differ from the flatter of the cube's sides by being more strongly curved, thus directing the entire shape upward. Or to realize that forelegs and back legs are bent at right angles and that this lion was carved out of a cube completely, or almost completely from memory. If the sculptor had had a recumbent or crouching real lion as a model, he could not have helped depicting the rotation between upper body and pelvis (391), because all quadrupeds rotate their pelvis as they lie down, so that one leg is under their body and only the paw or hoof is visible from the other side, while the other leg is stretched out in front of them. But even the lion crouching before it springs looks quite different, and never so rectangular. So what we have here is the simplest image schema, constructed of cubes. It is stored in the reproduction memory and has been given vitality through the basic processes of directed expansion upward and sideways, i.e. through the uneven curvatures of the edges and surfaces. The two cubes, displaced in a parallel manner between body and head, are connected by simple diagonals. Parallel hatchings, a frequently mentioned pattern of Hoffman's, faintly indicate the lion's mane.

So far we have confined our examples of ornamentation to applied art, to vessels, jewellery, and liturgical objects from the past millennia, because they could not be affected by technical or other considerations. If we had claimed that the rectangular cubic house or the individual building developed from a cylinder had come from the geometric Euclidean imagery of our visual cortex, it would have been easy to counter that the cubic house fitted best into the row of houses along the street; that this was the easiest way to plan its rooms without having pieces left over; that there were statical reasons for a cylindrical house as an individual object. Such arguments make no sense in the case of ornaments. It takes more effort to produce a precise chessboard pattern, concentric circles, radials, spirals, exact parallels than scribbles.

In depicting a lion or a human figure there is no reasonable explanation for its cubic form except the cubic imagery in the visual cortex. It is no easier to carve a rectangular than a diagonal or rotated block. On the contrary, most quarrystones are not rectangular. By observing nature, perhaps with some slight technical effort, it would be possible to create ungeometric forms from their fortuitous shapes. No, the reason is our visual memory. It is not easier to make, but to imagine the cube.

We construct the phenomena of our environment from simple geometric forms and store them in our visual memory. A first step in the direction of nature is the distortion of these geometric forms by means of the basic processes of directed expansion, in this case the suddenly increasing curvature of the side vaultings towards the edges of the original cube. Of course, these geometrically constructed forms are also based on the observation of nature, but geometrized. So the lion's whiskers are as evenly spaced as Hoffman and others' circle-radial "orbit", parallel hatching depicts its mane, its claws are identically long and wide. Ornaments have already made us familiar with such figures. Although the lion's eyes are not radially organized circles like those of the human idol from the coast of western Spain, 2000 BC, they are, despite their vaulted triangular shape, doubly rimmed, identically wide and deep, to surround a centre, and are thus developed by means of distortion, from the "orbit" of concentrically drawn circles. Real eyelids are unevenly shallow and deep in places.

From 3000 BC onwards Egyptian architecture, the palaces of the Thinite era or of King Zoser were constructed on strictly rectangular and cubic lines and their façades were structured by groups of identically wide parallels, like the corresponding Hoffman "orbit". This is demonstrated by the palace relief on the tombstone of King "Serpent" (in Paris) (380) and also by the relief resembling a palace that forms the sarcophagus of Ravêr (381). The Pharaonic symbol is developed from a circle. In addition to these we have the third, and best-known form in Egyptian architecture: the pyramid. These were built almost exclusively during the era of the Old Kingdom between 3000 BC to 2500 BC (the pyramids of Gizeh, 382). The entire arsenal of Egyptian architecture consists of cube, right angle, cylinder, parallels and pyramids. Of course, the simplest construction is the pyramid, statically like the principle of the heap of sand. But this no longer applies to the right-angled pillar temples (383).

The same geometric forms dominate Egyptian depictions of people (386, 387). The earliest, fairly natural female figure I know of is Egyptian, predynastic, Batari culture, second half of the fifth millennium BC (356), made of two cones set point to point on an ellipsoid base, with cylindrical legs, arms bent at right angles and the curved breasts unevenly directed. It was carved at a time when normally, all over the world, there existed only the simplest geometric idols, like the two bearded Egyptian figures from the first half of the fourth millennium or the French menhirs, 2000 BC etc., etc. If the date printed in the catalogue "Afrika" (Zeitgeist-Gesellschaft Berlin, exhibition from 1 March to 1 May 1996) is correct (the piece was lent by the Trustees of the British Museum), then the sculptor must have been a sort of genius of the fifth millennium.

Later, free-standing Egyptian figures were more dependent on the cube, for instance Prince Rahotep and his wife Nofret, c. 2800 BC (386, 387). They sit on cubic thrones, their legs are planted parallel, side-by-side, and are only slightly distorted cubes because of the slight vaultings of the uneven curvature and the openings between the legs and the feet. Nofret's body consists of two cubes, the pelvis and the upper body, widened by the arms. On arm is bent at right angles, the other almost so, as are the legs. The whole figure is strictly axially symmetrically constructed of horizontals and verticals, the only deviations being the sloping lines to the woman's breasts and to the man's thorax from his neck, and the slopes made by the tapering from the width of the pelvis to the knees. The thorax of the prince's figure also widens. But there is no doubt, essentially these figures, with trunk, legs, arms and head, are constructed of cubes, distorted by the basic processes of uneven swelling and curvature. I had already mentioned Queen Nofret, because of her circular radial necklace and her coronet. She also has circular breasts, their points, unlike real breasts, exactly central, very geometrized. I must also point out that Nofret's hair has parallel hatching, like one of the Hoffman "orbits".

There is no reason to say this is owing to lack of skill. The figures are painted, the colours have survived till today, The skilfully set eyes are made of quartz, surrounded by a setting which reproduces black eye-make-up. We are looking at high technical perfection. There are unlimited examples that can be so interpreted. Let me just also mention the almost life-size King Mykerinos, first half of the third millennium BC, with the two female deities of Upper and Lower Egypt (388). We see at a glance that the three figures are constructed cubically, that the dominant surfaces are horizontal or vertical. The right angle, the straight vanishing points, even the stretched arms obey the geometry of the right angle and the straight line (in reality, when an arm hangs down it is slightly bent). The dominant forms are the cube's planes, despite the surface, most sensitively distorted by the basic processes, which clearly indicates observation of nature. Things get really interesting at this point:

It is impossible to ignore the fact that we have here an actual portrait of the face of King Mykerinos, with his downward vaulted and directed royal headdress. If we look at the faces of other Egyptian rulers and officials (385, 384) we see that these are obviously the portraits of individuals. It is quite possible that the originals actually sat for the sculptors. Of course, all the parts of the body and the head itself are more cubic than in real life. A deliberate stylistic principle? Definitely not, for we do not just geometrically simplify our memories, we see that way, too. Here the two are combined. It is certain that these group and individual figures were created partly from memory and partly from the observation of nature, since both factors are connected by the type of our visual perception. Again we see parallel hatching on the hair, a horizontal, vertical or radial loincloth, the folds evenly spaced, the knees, in real life unsymmetrical, constructed almost symmetrically, but in contrast to the earlier example both women's breasts unevenly curved and so directed forward and downward.

Egyptian hieroglyphs (picture writing), too, like the carvings (389), consist almost exclusively of horizontals, verticals and diagonals. The same applies to reliefs, more freely interpreted, thanks to their narrative character, which are distinguished by an amazing dominance of straight lines, right angles, horizontals, verticals, sloping and parallel lines (390, 415). There is scarcely any art in the world in which the observation of nature and the dominant geometric forms of our visual cortex are so convincingly combined as in that of Egypt. Their world is re-created, completely from a geometric point of view, and yet seems natural. Such harmony between our brain's geometric inventions and the actual forms of the phenomena around us is certainly one of the reasons why this art could survive over 3000 years without becoming stereotypical or a just a copy of nature. That was reserved for the twentieth century.

We shall now turn to our next example, comparable in some ways, yet also quite different: Greek sculpture and painting.

Chapter 15
The Metamorphoses of Geometry in the Painting and Sculpture of Greece

The little, 20-centimetre-high, bronze figure of Apollo from Thebes, now in Boston, resembles the geometric type of human figure depicted on the shoulder zone of the amphora from Dipylon, 155 cm (402). He is clearly constructed of individual geometric or (392) geometry-like bodies: the – in cross-section – ellipsoid, cone-shaped upper body, which collides horizontally with the cylindrical legs which, seen from the front, bring pelvis and legs together to make one form[46]. From the back the cylindrical thigh-shapes are divided into almost circular buttocks and thighs. These geometric forms are slightly distorted by the basic processes. The thighs have a slight main curvature at the bottom and a strong main curvature at the top, towards the edges of the hips and pelvis. The upper arms are similarly curved. On the other hand, the geometric structure is maintained up to the head. The horizontal shoulders bear a long neck, formed from a stumpy cone, which in turn bears a head with a triangular face, made of an upside-down cone shape. The horizontal/vertical axes are so important that the central line of the abdomen (linea alba) is continued along the neck to the chin. The hair falls to the shoulders in evenly wide rosette-like compartments and has evenly wide parallel hatchings, familiar to us from so many examples. This massive division of the body at the waist, in the case of the vases and of this Apollo, is possibly traceable to the Cycladic-Greek idols of the fourth and third millennia BC, but may also be influenced by Egyptian reliefs (415).

We are not going to follow the developments step by step; we shall just look at important individual phases. The youth from Attica, now in New York, dates from some 70 years later (c. 620 BC). His thorax, abdomen, pelvis and thighs flow more naturally into one another (393), but the individual aspects of his form are still amazingly geometric, in contrast to nature. The curve of the thorax is pulled downward from the sternum to both sides in two halves resembling segments of a circle, equally deeply carved, and closes the thorax off from the abdomen in a linear manner and almost geometrically. The triangular counterbalance to the thorax aperture, the loins, goes from both pelvis edges, equally deeply carved, to the genitals. Even the knees, which in reality are absolutely unsymmetrical, are combined to form symmetrical ornaments made up of two curves, resembling segments of a circle, and a circular or vaulted triangular-shaped kneecap. The construction of the figure is strictly vertical-horizontal. The arms hang straight down, like those of the Egyptian figures. It is safe to assume that a south-eastern and Egyptian influence is at work here. The fluted pillar shafts, presumably adopted from the Old Kingdom, are another indication of this.

The construction of the back is also strictly symmetrical and geometric. That starts with the buttocks and the symmetrical edge of the pelvis and continues upward via the shoulder blades, with their quadricircular, equally deeply carved boundaries, and finishes at the hair, which is caught together in a completely screen-like plane surface, pulled back over the skull without reference to the

[46] Tonio Hölscher op. cit.

way hair really grows. The screen is interrupted by a totally symmetrically tied headband.

Despite this, the geometric forms on which this is now all based have been lent some vitality by the directed expansion. The upper arms (biceps and triceps) and the shoulder joints with the deltoids are directed by the upward pointing stronger vaultings, the main vaulting of the abdomen points downwards. In contrast to this, the main vaultings of the lower legs are directed upward through the way the calves are directed a little deeper on the inner surface and wider upward on the outer one. Obviously the basic geometric imagery has now been combined with observation of nature.

Nevertheless, the basic cubic structure remains dominant. The front, back and two side surfaces of the originally imagined cube are so slightly vaulted that only four main views of the figure are actually intended, and they collide at the vertically vaulted edges.

The figure still lacks almost any human movement of its own. The arms are just as straight and unnaturally stretched as those of the Egyptian figures. One foot is taking a small step forward, without, however, influencing the body's movement. Nevertheless man is beginning to move independently. There is a schematic indication that the legs can stride backward and forward.

Another big step has been taken towards naturalness and away from clear geometrization in the "Standing Youth" from Thenea near Corinth, c. 550 BC, (394) now in the Glyptothek in Munich. Even if the thorax aperture is still linearly carved like a parabola, at least it disappears more discreetly into the muscles at the side of the abdomen, the oblique and transverse ones. The main vaulting of the pelvis is directed upward to the waist, but is clearly harder towards the front point of the pelvis, getting softer again for the loins, the lower abdomen and the genitals. Nor are the arms still stretched as unnaturally straight as those of the Egyptian figures. They are slightly bent, so that the elbows are behind the waist, and the lower arms and hands intersect the thighs in front. The knees are carved more naturally, no longer symmetrical. The thigh muscles end pulled down inwards unsymmetrically around the triangularly vaulted kneecaps. The cross-section of the entire thorax is much more vaulted, both at the back and front. Shoulders and arms are no longer almost on one level as they were in the case of the New York youth's pectoral halves, carved in a plane-surface, but the chest muscle is curved around the thorax. Even though the hair is still formed of identically wide rolls, it changes shape as it crosses the skull and combines better to make a whole. It no longer seems to consist of separate units. But this youth, too, is only taking a hesitant step forward – however, the position of his arms, now that they are no longer rigid, is far more natural and complies with the fact that in the human arm the flexors are stronger than the extensors. The volume of the face, with the cheeks, the corners of the mouth and the cleft chin, is far more strongly distorted and thus combined with the slightly slanting eyes, the nose and the mouth to a mobile plastic unit. Man is starting to smile.

Now, if not before, some will be raising the objection: Why are you calling this geometrized? This is what real people look like. But a glance at the photographs of the nude men (448, 449) shows, without more detailed explanation, what a great difference there is. This stone figure's swelling life is far more geometric than in the case of the real men and, for us, this figure gains vitality thereby, because we sympathize with this struggle between geometry and basic processes and because the one intensifies the other.

Before continuing along this line, let us first deal in a similar manner with the kore, the clothed maiden.

The oldest surviving life-size Greek marble statue is the Nikandre of Delos, 1.75 metres high. Basically she is just a cube, hardly distorted (396). Her legs are parallel, her breasts hardly distort the front of her thorax. Abdomen and legs are only separated from thorax, arms and head by a slight dip above the hips. Although the figure is badly damaged it is clear that there are no pleats in the skirt, which, however, unlike those of many Egyptian figures, is not modelled to the contours of the body. The straight, almost rigid arms hang down axially-symmetrically at the sides of the body. The hair, falling to the shoulders, is divided into two blocks of equally wide locks, framing a triangular face down to the triangle of the neck. The Egyptian influence seems to have been more or less misunderstood here. After all, the late third-millennium idols, even though much smaller (340, 341) already had far more plastic qualities, were more differentiated because of the distortions by the basic processes, but here, c. 750 BC, we have a geometric – almost block-like figure – a primitive depiction of a human being – at the end of the so-called dark ages which roughly started with the war between the Hittites and the Greeks (better known as the Trojan War).

We now leap forward 150 years to the high archaic female figure with a pomegranate from Keratea in Attica, now in Berlin (397). Even though the construction of the whole figure is still strictly axial, the two legs are side by side in a block, the middle line of the pleated skirt and the space between the sides of the mantle hanging over her shoulders account for her upright head, so that one can say that the middle line is continued from the nose – the arms are no longer straight or at right angles; the arm with the hand holding the pomegranate is at an obtuse angle, the other one at an acute angle – and more direct, plastic vitality has come about through the construction of the face. Eyes, nose and mouth have not just been set horizontally-vertically on a base; they combine on the one hand with the mobile volume of the features, with cheeks, upper lip and chin, and on the other hand, through the abandoning of the rigid right angle, there are slanting lines and even the beginnings of radials, creating a smile. Although the hair is still in identically wide locks, it now slants from a central axis, perhaps even from roots hidden under the hat. It is no longer a strictly rectangular screen.

But the start of a vitality which can overcome simple screen and cube geometry is still combined with the strict geometric parallels of the identically wide pleats of the chiton. The only folds which are much narrower are those of the mantle over her shoulders. However, the whole appearance is reminiscent of the parallel Lie "orbit", or, to draw on a comparison from the history of art, of the architecture of the Old Kingdom and the Fifth Egyptian Dynasty, organized in the manner of reliefs. There is a direct resemblance between the reliefs on the sarcophagus of Ravêr from Cairo (381) and the robe of the female figure with a pomegranate (397). While in the case of the sarcophagus the different widths of the parallel and flat surfaces of the groups of stripes alternate, in the front view of the female figure the width of the pleats alternates between the mantle at the top and the chiton at the bottom, but the pleats of both pieces of clothing are identically wide and almost completely parallel. The areas in the front and down between the pleats come together without any rotation to form a single surface, repeating the top surface of the bodily volume, thereby differing not at all from the palace façade of the Old Kingdom. From the back the pleats of the chiton and, swung round horizontally to them, the pleats of the mantle are even the same width, but the slanting stripes of hair are much narrower and so form a contrast.

It would be totally misleading to assume that the sculptor in Keratea, Attica, around 580 BC had seen the sarcophagus of the Pharaoh Ravêr from the Fifth Dynasty or any of the palace architecture of the Old Kingdom (399), but the other way round the whole thing makes sense. Both the architecture of the Old Kingdom and the pleats on the figure in Berlin are based on completely parallel lines and identically wide spaces, or the "schematism of simplicity". The brain's perceptual images are the same; what they express can be very different. Radial "orbits" can be used for the infant's tadpole, for the eyes of the idol, c. 2000 BC, from western Spain (336) and for the shape of the necklace and the headband-decoration of Nofret, the consort of Rahotep, or to symbolize the nipples of Queen Teo of Egypt (400).

The continually changing meanings of the same patterns – spirals, for instance, can symbolize water, sun, fertility – indicate that these patterns are based on the same geometric ideas but have not spread across the world by means of cultural export or import. In that case they would have taken their meaning along with them. They have a cerebral origin. They arose everywhere independently of one another. They always symbolize whatever happened to be interesting at the time or whatever they seemed to best fitted to represent.

Even if the Hera of Cheramyes (found in the shrine of Hera on Samos, height 1.92 m) is not made from a cube (398) but grows upward like a column out of a rather strictly maintained cylinder, the entire modelling is much softer than that of the figure we have just considered, but the legs are still like two side-by-side pillars and the chiton still has identically wide parallel pleats. At the bottom, in a negative curve, it fits in with the circular plinth. Yet again we see the "orbit" of the parallel lines and the identical spaces. This is also essentially true of the mantle; its pleats are also parallel and almost identically wide, but now fall in a generous, unevenly curved line from the right shoulder to the hips. One arm is straight, in the strict Egyptian manner, the other is bent and lies across the heart. The pleats of the mantle, running diagonally and parallel, are unaffected by the thrust of the breasts; at least, there is no observable difference in the way the pleats fall. Despite the diagonal folds, the mantle finishes in a smooth, geometric, parabola-like line above the chiton. It takes no notice of the fact that, because of the folds, such a simple geometric finish would be impossible. The elements of this column-like figure are: circle, cylinder, vertical parallel lines and the uneven parallel curves of the folds of the mantle, the bottom edge of the mantle, the formation of the back and buttocks. All the folds are parallel surfaces, with no rotation.

Another big step forward to the end of the archaic period around 500 BC is represented by the girl in the Acropolis Museum wearing a chiton with sleeves and a mantle (401). Wavy lines form both the top of her chiton and her hair. We have seen these as ornamentation and in children's drawings. They are parallel in front and radially organized from the shoulders to the armpits. Her hair slants from a parting and is held by a diadem-like headband. The softly sloping shoulders and the arms, or what has survived of them, are modelled in a soft organic fashion, as is the face. The archaic smile has been transformed into a calm Mona-Lisa-like expression, but, unevenly curved, assembles all the features more softly and gently so that their organization is both horizont-al-vertical and radial. The slanting lines of the eyes, cheeks and the corners of the mouth form a radial structure which combines with the horizontal-vertical structure of the nose, mouth and forehead to create a calm, mild expression.

The mantle demonstrates a really important difference. From the right shoulder it is slung in a generous curve under the left armpit and from there falls downward in wide folds, which are no longer strictly parallel. They vary in width and – most importantly – have rotated surfaces. With their main surfaces the folds incline towards or away from each other. Follow the course of the folds' surfaces from the centre line to her right arm and her right armpit. As the rotated surfaces and widths change, parallel lines are abandoned. The folds diverge and come together under her arm. The hem, seen against the chiton, is no longer a geometric parabola or any other Euclidean form, but is formed by the uneven curves of the mantle's many folds. Only the top frill from the right shoulder to the left armpit forms an almost evenly curved segment of a circle. It appears to lie over a cord and forms a running spiral of the kind we have seen ornamenting vases.

We can still recognize the basic geometric forms, e.g. the regular up and down of the wavy line; the spirals; the parallel vertical; the edge of the hair against the face, made up of two circular segments resembling a symmetrical Gothic arch, but they are partly distorted by the basic processes of the directed expansion of the rotated surfaces and the uneven curvature and vaulting, giving rise to a harmonious balance between geometry and nature. Awakening vitality is symbolized by the metamorphoses of geometry. This expression has an even stronger effect on us precisely because this figure is not a copy of a real person, because, despite its nearness to nature, it is not naturalistic but is a new creation, made up of our visual cortex's geometric concepts and the metamorphoses of this geometry. It originated in our minds and is thus not only a copy of nature but also a product of our imagination. The metamorphosis of geometry into a natural form, or the regulating of a natural form into a geometrical one, creates a universal picture of man from a fortuitous phenomenon.

But this figure (401) also raises the question of human or, in other cases, animal movement. If we imagine the rest of the girl's arms, we can see that they were bent at different angles, the axis of her head and neck is no longer subject to the rectangular schema, as her head inclines slightly to the side and forward. If we complete the figure in our imagination, one leg seems to be in front, or even to be the non-supporting leg. The young girl appears to breathe, not only because the basic processes have transformed the geometry, but also because truly human movement has become the theme of the depiction.

So now we come not only to the question of the basic processes which create life, growth and decay together with spatiality, but also to the question of the perception of animal and human movement and our memory of it.

Chapter 16
Movement Schemata

The simplest human and animal geometric form schemata available to our visual memory – such as the examples in the drawings of humans and animals by children and adults – give no hint of the specific mobility of these beings. The most we can gather from the child's drawing of father, mother and child (42) is that these people are standing upright on their feet – although even that is not completely clear – just as we cannot detect any movement in the depictions of people and animals on the North American bag (351), despite the far richer geometric details. 80% to 90% of the drawings of elephants and cows by children and students, with the rectangular, parallel legs lacking detail and meeting the bodies at right angles, give no hint of how these animals move. We have already discovered this in the case of the growth movements of spruce, palm and fern. Nor do the plastic animal depictions of the Greeks' proto-geometric period, of the Hallstein period, of the very early Egyptians, of 1000 BC and earlier in China give any idea of the animals' specific manner of movement, with the possible exception of the elephant's swinging trunk.

So the simplest human and animal geometric form schemata in our visual memory give barely a hint of the mobility of the beings depicted. Of course there were no children or students unaware of the fact that animals and people set one leg in front of the other, turn and turn about, in order to move. But unless this problem is explicitly mentioned, they tend to reproduce form schemata which do not even hint at such mobility. In some of the elephants drawn by the 11 to 13-year-old children there is at least an indication that the elephant can walk (285, 287). Only two legs meet the rectangular or almost rectangular body at right angles, while the other two slant and thus intersect. When we come to the students, only 12.6% drew elephants whose four legs "moved" in this fashion. The cows fared a little better: although only 15.3% of the students depicted a recognizable step by drawing sloping legs, 25.5% depicted a certain difference between the forelegs and the back legs, admittedly without indicating any to-and-fro movement.

Most quadrupeds, whether we are considering cows, horses, dogs or any others, resemble each other in that usually the back legs provide the forward thrust and are thus more strongly formed down to the heel and then thin down to the hoof or paw, while the forelegs, although also set one before the other, are rather responsible for supporting the body. As we see, the simplest purely geometric schema does not contain to-and-fro leg movement because the identically rectangularly formed forelegs and back legs have to meet the body at right angles. Movement cannot be depicted until we introduce diverse formation, the strong bend in the back legs and the powerful development of the lower leg. Strangely enough, a quarter of the students bent the forelegs the wrong way. That undoubtedly indicates not a lack of drawing ability but severe faults in the visual memory, because it is certainly just as easy to draw an angle pointing backward as forward. So, basically, in order to reproduce even the simplest movement, such as walking forward, the Euclidean geometry of the form schema has to be distorted, or, to put it another way, the metamorphoses of geometry are a precondition for committing the course of individual creatures' movements fairly accurately to memory and for depicting or describing them at least roughly.

At the beginning of the second part of my book I pointed out that most of the movements of the higher mammals, but also of reptiles, birds and many other creatures, are so complicated that we cannot even follow their actual course, let alone remember it. I compared stills taken of horses with the way horses were depicted in art before the invention of stills (278). Pictures painted before the invention of photography show what the artist could actually remember. Paolo Ucello's movement schemata, for instance, are so reminiscent of a rocking-horse's that they are almost comic.

Our imagination's movement schemata are derived from each other in the direction of an increasing individualization and complication or, with the increasing universality of their meaning, set above each other. Thus we have a very general idea of the way quadrupeds walk, confined essentially to their setting one leg in front of the other while body, head and tail move evenly horizontally forward, a kind of car on legs. This very general movement schema is still very close to the basic process of horizontal displacement. Even so, it made no appearance in 60% to 80% of the drawings by children and students because it can be only imperfectly depicted using rectangular legs or bodies and because the people making the drawings are obviously hardly thinking about the course of the movements, not even the movements of the legs, and have only a vague idea of them in their visual memory. The more complicated such a schema becomes, the more an animal's walk appears to unite the movements of body, neck and legs. That is certainly the case with Paolo Ucello, who sometimes combines them almost grotesquely; for instance, the horse on the right, kicking up its back legs (278). But even the schemata our reproduction memory can manage to memorize have been very simplified when compared to animals' natural movements.

We can find the same subordination and superiority of general and individual movement schemata of our visual memory which approach natural movements for all the life forms known to us, even for human movements.

But even when the geometry is already slightly distorted, as it is on the sepulchral vase of Dipylon, c. 800 BC (402), where the horses' back legs look somewhat different from their forelegs and even the people's legs are divided into thigh, knee and calf – although the people's bodies still consist of two triangles balancing point to point and the horses' bodies are flat rectangles – even if the form of the animal and human legs fulfils the preconditions for walking forward, the "orbits" of the parallels are so dominant that we cannot make out any such movement, i.e. a deviation from the parallels.

The right angle, particularly when horizontal and vertical, still dominates the imagination to such an extent that the chariot-wheels' cruciform spokes are all horizontal and vertical. We do not get any impression of turning wheels. Immobile people and animals, rigidly immobile wheels, even though the scene is supposed to depict funeral games with war-chariots in honour of the dead.

It is no wonder that the decoration of this vase still consists of the chessboard pattern, the concentric circles of the "orbits" and radials, the parallel straight and wavy lines. Even the arms of the mourners are all lifted and bent at right angles, so that the head of each mourner is always in the centre of a rectangle (402a). Despite this schematism of simplicity, confined essentially to rectangle, triangle and circle with radials and parallels, the total impression is one of a rich and varied ornamentation.

In the same era we find the same stage of completely schematized movement, confined by rectangular geometry, in sculpture. An example of this is the man confronting a centaur, between 800 and 700 BC (403). Admittedly, at the back the centaur has equine legs, as we see from the heels and strong lower legs which, however, combine with the gaskins and hindquarters to form an upward swelling vase-like shape. The forelegs resemble human legs, but he and his opponent are completely locked in a rectangular schema of the kind that dominates African sculpture (350). Even the confrontation, the arm wrestling, is frozen at a right angle, the distances all almost the same.

About one hundred years later striding is clearly represented in the relief "Kaineus and the centaurs" (404). However, since all the strides are nearly identical; only a large and smaller space regularly alternate between the horses' and human feet; the centaurs' arms, although no longer exactly horizontal/vertical, are bent almost at right angles; in contrast, the angle of Kaineus's arms is obtuse, though they are almost symmetrical; again, on the other hand, the two swords are almost at right angles to the forearms, so that Kaineus represents the exact symmetrical central axis of the picture, which is supposed to be mobile – all this gives rise to a very simple movement schema, which is by no means the depiction of a dramatic struggle. Striding legs and bent arms are not enough; it would completely distort all the bodies.

In the equestrian games on the Corinthian colonate crater from the beginning of the sixth century (405) we see galloping horses, their back legs stretched far backward and their forelegs stretched far forward, in much the same way that this is often seen in Chinese painting (280). But the movement is still strongly schematized; there is no contrast to these huge strides: for instance, the four hooves bunched together or intersecting back legs and forelegs. The figures of the riders, the necks and heads of the horses, are all still strictly parallel. What we have here is a jumping schema, an advance on the preceding example, but still a movement schema. There can be no doubt that our visual memory works not only with form schemata, but also with simplified movement schemata.

On the bowls of the Penthesilea painter and of Euphronius (406, 408) the horses' movements are far closer to nature. We can see head and neck movement and complicated leg positioning. But it is impossible to ignore the distortion, which is still relatively close to the simple geometric form. As a general rule the development of the depiction of movement happens faster in painting than in sculpture, even if up to the end of the archaic period rotated movement (see the crater of Ergotimos and the painter Klitias) (407) is actually still very rare. Profile, full-face or three-quarter profile – in contrast to the body only the head is rotated, but we do occasionally find the beginnings of body rotation between the pelvis and the thorax – still rather clumsy – in the transitional period between late archaic and early classic, for instance on the crater of Euphronius (412). Even the forms which are close to nature, such as buttocks, thighs, lower legs, chest muscles and pleats, when compared to nature have all been geometrized. The uneven curvatures are smoother and their geometric origin clearer than in nature (409).

But even in the pictures on vases, c. 500 BC, in which forms and movements that are close to nature are portrayed, e. g. (409 a, b, c) "Achilles tending the wounded Patroklos" or the "Athene" of the Andokides painter and his "Hercules", the inner drawing is purely geometrical – not distorted in perspective. Achilles's and Patroklos's armament consist of different types of purely undistorted screens. This is true of the "Athene" of the Andokides painter as well as Hercules's spiral locks and the radial pattern on his material. This remark is almost always true of vase painting until 450 BC.

The late archaic period offers some examples of such movement schemata, for instance the Aphaia temple in Ægina, c. 500 BC (410). Just as, in my opinion, our visual cortex's simple Euclidean geometry is essentially two-dimensional, so are the form schemata two-dimensional. It is not for nothing that the Attic youth, 620 BC, in the Metropolitan Museum has four aspects (393). The frontal and rear views are combined in a block with the two profile views. But they only partly form a unified volume which also contains circle-like or even spiralling processes.

In May 2000 we asked our students (there were some 90 students between the 6th and 15th semester present at this lecture) to draw a running or walking person in profile. 10% at most of the drawings showed a figure which – quite apart from the individual's drawing talent – in profile indicated by the forward slope of the trunk that the person was actually running. But even in the case of the best three figures not only the details were very geometrized; in addition, as a rule, both arms were angled identically, usually at right angles or slightly obtuse, and one leg was also at a right angle, sometimes both were (416 – 418). Since in the case of the three best the trunk not only sloped but consisted of its three individual parts, i.e. thorax, pelvis and abdomen, the leg and arm movement was comprehensible. It must be said that these three people had taken part in my voluntary life class and so had gained a little knowledge and some ability to see. For the rest of the 10% the trunk was a unit (419 – 422) which, even if not always rectangular, was not distorted by the arm and leg movement. However, if the trunk was drawn leaning forward, there was a schematic impression of running, although nearly all the legs and arms were angled identically. But between 60% and 80% of the students did not even manage a schema indicating that the figure was supposed to be running forward. (It is difficult to give precise percentages, because the transitional stages slide into one another.) (423, 424, 426, 427, 428, 430)

Now it would be possible to argue that the three best students had attended my life class and so could draw better. But that is not completely right: at this point all architecture students must have attended many hours of drawing class for four semesters under another professor. (My voluntary life class, in contrast, lasts only two semesters with three lessons a week.) The only reason for the big difference in the quality of the drawings is the difference in visual knowledge of the human body and thus a different quality of reproduction memory. Anyone who thinks the main problem in drawing a runner is a lack of drawing ability should check this by trying to draw such a figure from memory. Knowledge and memory are the problems, not technical ability. Our three students prove how important drawing lessons based on knowledge are for visual perception and visual memory. We shall come back to this point.

The running warrior from the eastern pediment of the Aphaia temple in Ægine, c. 500 BC, is a classic example (410) at an advanced stage, if compared to purely geometric depiction, of such a movement schema. The individual body parts are clearly separated and fairly close to nature. However, the artist has drawn the same obtuse angle for both arms and the front leg. Only the right leg is stretched out, thrusting the figure forward. But when someone walks or runs, his trunk rotates slightly around his spine, which is why his arms swing, but here all the levels are parallel. If we join up the hip bones, the two chest muscles and the forehead, we get parallel areas, by no means rotated ones. There is no trace of the constant rotation caused by running. The somewhat earlier running warrior from the frieze around an Attic clay vessel, 600 BC, has more rotation. While the pelvis is shown in three-quarter profile, we have almost a back view of the shoulders and we see the head in profile (411). So the classic schema of a runner (410), somewhat closer to nature, is a sloping body with arms and legs at identical angles and an unrotated trunk.

I have already pointed out that this rotation between pelvis and thorax was not depicted in the sculpture of the recumbent Egyptian lion and the lion-panther from the western pediment of the Artemis temple, 600 BC, sits in the same way, on back legs and forelegs bent vertically at right-angles. (413) Even if the individual forms are more curved, swell more than did the Egyptian example, they are still only geometric forms simply distorted by the basic processes. The relief, despite the rotated head, makes a two-dimensional impression. I feel I should also point out that the many Lie circular "orbits", applied as a regular screen, are probably intended to represent the fur and that simple parallels represent the ribs. Even the four bronze lions of the famous seven-branched candelabra in Braunschweig Cathedral (twelfth century) are cubic and right-angled.

A classic movement schema from the western pediment of the Artemis temple on Kerkyra is the "Dancing Gorgo" in the centre of the pediment (414). Here we are apparently looking at a rotation between pelvis and thorax. We have a full-frontal thorax showing both breasts, while we can see an unevenly curved buttock protruding from the pelvis, sloping downward; i.e. the pelvis is depicted in profile. But the whole thing is put together rather like the Egyptian reliefs (415), there is no actual rotation. Instead her four extremities (as far as they are visible), legs and arms, are all bent at almost the same right angle. With the exception of her left arm they all deviate only slightly from the horizontal and vertical.

As we see, movement schemata and form schemata are almost indistinguishable. They are synthetic perceptual images. The pure Euclidean geometric form is rigid and immobile, but as soon as it is distorted by the basic processes and has a direction it can depict schematized movement. Even in the case of the "orbits" it is not clear whether the intention is to show the basic processes or the geometric forms. The two things overlap. The form schemata are constructed of Euclidean forms and distorted by the basic processes of rising, falling and directed expansion, which brings them close to their natural models. But there is also the beginning of intensive observation of nature, leading in classic and Hellenistic times to figures we now generally dub "naturalistic". Only a very few people can spot the difference between them and actual nature, because we no longer know how to see and have forgotten our visual knowledge.

Before proceeding we must turn once more to the three types of movement. It is not very important to decide whether we are dealing with an actual alteration in time and space or only an imaginary movement. The form of two categories overlaps, as everyone knows who has seen fruit, leaves or trees growing. We have already discussed this point. Of course, it also applies to the movement schemata, which are really only imaginary. Real beings move naturally, not schematically. But we cannot only draw, paint or carve our very simplified ideas about the process of such movements, we can use marionettes or other puppets, animated cartoons etc. to depict such movements. Part of the reason why we understand the movements of these marionettes and puppets so easily is that they correspond with the movement schemata stored in our memory. For this reason we shall not bother with the difference between real movement in time and space and the movement conveyed by the individual form.

A rising, unevenly curved vase, for instance the Egyptian one, end of the fourth millennium BC, (64) shows us an upward directed swelling expansion, getting larger – the vase seems to be quietly expanding – though of course it is not doing so. We are looking at a sphere, distorted by an imagined basic process of directed expansion. We compare this uneven curvature with even curvature and the difference indicates growth to us, as we have seen it in fruit, leaves and all the other examples we have considered.

So we see that the basic processes have a remarkable double character. On the one hand they are independent of any meaning. I can imagine expansion affecting a whole space or just a piece of fruit, horizontal displacement or a rising and falling movement without attaching any particular meaning to it. I do not have to imagine the Big Bang, which is said to have produced the universe, or a growing apple, an unfurling leaf, an unevenly curved vessel; I can think about such movement without attaching just one meaning to it. In this way it differs from the movement schemata and natural movement. I cannot imagine the movement schema of a human being, a horse, a cow or a bird without thinking of the creature in question, and I certainly cannot imagine its natural movements without thinking about it. They are precisely fixed in their meaning. So movement schemata or form schemata distorted by them and natural forms and movements are synthetic phenomena, while geometric forms and the basic processes can also be pure perceptual images, because we can imagine them without the other matching perceptual image or memory picture.

On the other hand there is no living creature in the world, no natural phenomenon on earth, which is not distorted by the basic processes. A hilly landscape, all the vegetation, every living creature – everything expresses expansion or collapse, rising or falling, or horizontal displacement. It makes no difference what we are considering: fruit, trees, bushes, leaves or blossom; people and animals; or rain, snow, wind, a sandy beach; or rocks and icebergs. The basic processes are omnipresent in their forming power. They are one of the reasons for the infinite number of phenomena around us and that is why they are not indissolubly coupled with any single one of them in our imagination, in the way that the human movement schema is inextricably bound up with humans.

If we are watching Mickey Mouse using human schema movements in a cartoon, we find ourselves thinking of him as a person. The movement schema is so inextricably bound up with one single meaning that transferred to another creature, it changes that creature's essential meaning. But all these forms, whether artificial of natural, can only seem to be alive if they are directed and distorted by the basic processes. The basic processes are omnipresent in life.

Despite this, the movement schemata are different from natural movements, which we are frequently unable to perceive correctly. Everyone can remember how fascinating it is to watch a slow-motion film of a wild animal running fast or a sportsman. What a difference there is between the natural movements suddenly made visible and our schema imaginings. We are nevertheless only able to store in our memory some of the movements we have just seen in slow motion – and we can just do this for a short time. We can remember only a geometrized version of natural forms – simpler or more complicated, depending on our talent, the state of our knowledge and our experience. Our memories of natural movements are just as simplified.

Many will argue against this conclusion. After all, actors are amazingly good at copying the movements of other people or even of, for example, an orang-utan. But I say that that is the same process as in fine art: it is still schematization; it has just become more complicated through observation, experience and learning. It was not very difficult to imitate Hitler. A few typical observations combined with normal movement schemata or the actor's movements were enough to call up the necessary associations which, in reality, were a long way from his actual movement process.

So the basic geometric forms developed in our imagination from the basic processes. We recognize the processes of growing, maturing and decaying by observing the differences between nature's natural phenomena, distorted by the basic processes, and our geometric images. If we meet the geometric forms in our environment, if we produce them in an undistorted manner, simply as boxes, spheres or screens, they contain no information for our recognition, over and above the fact that they exist. That is why nearly all modern architecture is so cold. Only when we make comparisons and note the difference to our brain's geometric ideas do we recognize the processes, feel the life.

More or less the same applies to the natural movements of living beings. As a rule we cannot perceive them individually, nor remember them precisely. We simply store schematized, variously complicated versions of them in our visual memory. How good we get at this depends on our willingness to observe, to learn and to employ our knowledge.

Chapter 17
And Once Again the Visual Memory

My basic belief is that it is the reproduction memory as a part of the general declaration memory which makes us different from the animals. They have essentially only an identification memory – and perhaps a short-term memory. This reproduction memory is one of the most important bases of our consciousness, of visual perception, our ability to observe, in other words of thinking. The incredible way in which we currently neglect the improvement of our reproduction memory is one of the most important reasons for the increasing failure to think independently, to have become dependent on the media and to do the same things all over the world – and to neglect the same things.

If there are still perceptual psychologists – and there are – who claim that for all practical purposes there is no difference between our visual memory and our actual visual perception; that it makes no difference whether we actually see an elephant or only imagine one; that we can remember every detail of our neighbour's face – this is striking proof either that these psychologists cannot differentiate between the reproduction and the identification memory or that they have never put this question to the test. In fact the opposite is true. We have already spoken about the elephant, and seen the drawings: completely schematic, constructed entirely of geometric forms, not even a trace of movement. And our neighbours?

I really wanted to know the answer and so on Friday, 12 May 2000 I asked my students to draw from memory a person from their immediate family, someone they had known from day one, as it were: their father, from the front and the side. Looking at these "portraits" (432 – 439) we have to say that these 90 students have only two types of father: a father with or without a beard. From the front his face is more or less oval, his profile is almost vertical, with a long, protruding nose and often a mouth and chin area that is too short. The nose runs smoothly from top to bottom, with no difference between bony part and cartilage. Both from the front and side the drawing is almost totally flat – and that from architecture students between their sixth and eighteenth semester, who had all had to attend four semesters of drawing lessons with another professor. The beards of some of the bearded fathers make them look almost square from the front. Apparently the primary optical memory concerns the cut of the father's beard and whether he wears glasses. If, for instance, we were to assimilate the beard of the father in figure 437 with the beard of the father in figure 435 there would be almost no more difference between them.

A vertical profile is geometrically the easiest to draw. There is no memory of the fact that there are angled profiles with a receding forehead and receding chin or that profiles frequently develop from a sort of ovoid form or that the lip area protrudes as in a beast of prey – no, every father's profile is constructed from the vertical. We do not need photographs of the fathers to prove that the drawings resemble each other far more closely than the actual fathers do. What we have here are some variations on the simplest facial schema, such as we find over and over again in the masks and heads of Africa – with the exception of the Nigerian heads from the twelfth to the sixteenth century (468, 469). The schema is determined through the too long forehead-nose unit and the too short mouth-chin area. Although such an African schema can be very expressive, it cannot be characteristic or resemble a particular person. Students who model clay heads from memory follow this schema again and again, though mostly without expression. The heads produced by kindergarten pupils also look like this.

Once again, only three students drew differently (440, 441, 442). These were again the three among the 90 students who had attended my life classes, held in accordance with the methods I have described. The heads are three-dimensional, not flat. We can see the difference between the noses' bone and cartilage. There are cheekbones. The proportions of forehead, nose, mouth and chin to each other are correct. Now we really would have to consult photographs to determine whether these drawings are good likenesses. This is difficult to believe in the case of the first drawing (440). The father looks too young, about the same age as the student. With the other two (441, 442) it is a distinct possibility. They look quite like portraits. They are still simplified, but not simply schematic. Of course, students with such contents in their visual memory see differently and more precisely and can think more creatively than others. But our educational politicians, who rely only on modern techniques, no longer grasp this. There could be an important task here for the psychology of perception: to influence the politics of education.

Chapter 18
So-called Naturalism

Now if it is really true that our perceptual memory, long or short-term, unsupported by a visual confrontation with the imagined object, is based only on more complex or less complex geometric forms, we are immediately faced with the question: is our visual perception also geometrically simplified? In the case of spontaneous visual perception, not called in question by precise visual study, my experience as a teacher says: yes. But, some people will say, we perceive naturalistically, exactly as things are. After all, there was naturalistic art in antiquity, in the Renaissance and at other times. Very well, let us compare so-called naturalistic art – made by human beings, not the result of a photograph or a cast from nature, which despite the natural model is based on the reproduction memory and the short-term memory necessary for modelling the form – with corresponding natural photographs. Let us start with two Greek sculptures from the late classic period, which our contemporaries would certainly consider extremely naturalistic, and compare them with photographs of nude figures.

A Greek Perseus, now in the National Museum in Athens (found in the sea of Andikithira), based on the Polykletian tradition, is in many ways typical of the Greek classic period and early Hellenism. I also draw your attention to the almost intact marble statue of the Agias of Pharsolos, a masterpiece by Lysippos from almost the same time (446, 447). The two nude photographs, which are very different in their anatomical details (448, 449), show the much more ungeometrically flowing forms of a natural nude. This starts with the division of thorax, abdomen, pelvis and thighs. Let us consider the lateral muscles of the trunk. Whether a leg is taking the weight or not, the Greeks usually make the muscles above it at the edge of the pelvis stand out, in order to obtain a clear geometric division between the soft flesh and harder pelvis edge. In the photographs we see that the muscles bulge above the pelvis edge on the side of the leg taking the man's weight, but on the other side they are smoothed out, or at least run more smoothly from the thorax to the pelvis. The Greek depiction is not only unnaturalistic, but even illogical. In a counterpoise the pelvis sinks on the side of the leg not bearing the weight and the thorax rises, so that the distance between thorax and pelvis increases, while it decreases on the other side. Hence the bulging of the soft fleshy parts on the side of the leg bearing the weight, while on the other side they, or the muscles, are more smoothed out. The Greeks presumably ignored this in order to obtain a clearer, more geometric and cubic division between thorax, abdomen and pelvis, as already shown in the simple form in the small figure of Apollo (392)[47].

On the two Greek statues the linea alba (the vertical central line of the abdomen) is combined with the diagonal of the rectus (the straight frontal abdominal muscle) and the prominent forms of the thorax aperture to a course of smaller and larger vaulted-tense cubes. These are not only more geometric than those in the photographs, but also show a clearer contrast between small and large forms.

This more cubic construction of the Greek sculptures is particularly clear in their thorax apertures. In the photograph of the slimmer nude it runs from a point downward below the sternum and then disappears relatively vaguely between the thorax and the abdominal muscles. On the right-hand side of the other photograph the thorax aperture somewhat resembles those in the Greek sculptures. It is not so definitely cubic, but disappears rather vaguely diagonally into the straight abdominal muscles on the left-hand side when compared to the right-hand side. In both the nudes and the bronze sculptures the strongest rib in the thorax is prominent. However, while in the case of the Greeks this leads to a combination of clear, tense, geometry-like forms, in the case of the nudes these surfaces develop far less clearly, and this includes the composition of large and small forms. This also applies to the transition from abdomen to thighs. In the nudes this transition is more flowing on the side of the leg not taking the man's weight, while in the Greek sculptures the lower abdomen is finished geometrically with something resembling a semi-circle above both legs. The sculptured pubic hair is much more geometric and compressed than it is in real life.

[47] Tonio Hölscher op. cit.

We could continue this description by comparing all the details with each other, but we will leave that to the reader and observer. We should just like to add that in the Greek sculptures it is clear that the total structure of the trunk, despite all the details and the displacement through movement, is based on a more cubic structure. Something still remains of the idea of a front, side and back view of a cube in the distortion to bring it close to nature, similar to what we saw very clearly in the early archaic and Egyptian sculpture. With the real nudes, the transitions from front, side and back views are more flowing, vaguer and less geometric. To quote Dürer, "Nature lies in art, we only have to draw it out". I think he meant what I have just described.

The high Renaissance and Dürer himself were particularly influenced by Greek-Hellenistic sculpture and the Roman copies. I should like to draw your attention to Dürer's copperplate engraving "Adam and Eve", 1504 (454), particularly to his Adam, whose composition somewhat resembles in both movement and proportion the Greek Perseus statue we have just been considering. I have just analyzed this sculpture's cubic connections. In Dürer's pre-drawing for this "Adam and Eve" engraving in the Albertina in Vienna we can see that he based it on circles and cubes (456). We see that the special form of the thorax aperture in Greek sculpture, together with the groin and the lateral abdominal muscles more or less form a circle, which both Dürer and the Greeks centre on the navel (446, 456). There are also similar pre-drawings for "Eve", constructed from geometric circles, but also from rectangles or cubes with their natural distortions (455, 457). So we see that in the case of "Adam and Eve" Dürer proceeded from the same basic geometric forms, which he distorted by means of the basic processes and human movement. It is a matter of general knowledge that Leonardo da Vinci made similar drawings.

It is quite obvious that the Renaissance artists sought harmony of form in a combination of geometry and nature. Probably they came upon similar connections when examining models from antiquity. But we shall return to this later. For the time being we should just note that for the most part the picture compositions of the Renaissance and the high Renaissance were also composed of geometric forms. Art historians have emphasized this point so we need not illustrate it. Here, too, we can state as a general rule, that structures composed of only slightly altered geometric forms, such as we frequently find with Andrea del Sarto, provide less information and have less expression than geometric structures which have been distorted by basic processes and human movements. We can clearly see the metamorphoses of geometric forms into natural ones in the "Dionysos" by Phidias in the east pediment of the Parthenon (450). Although the effect of the body is heavy, soft, completely human, its genesis from a cube is not only obvious in the thorax, the fleshy parts of the abdomen and the thighs – it is the predominant theme: marble became flesh; geometry became corporeal form. Looking at the knees we can of course see the attachment of the flexor tendons to the calves, and yet we know that natural thighs (451) can never look as cubic as those of the Parthenon's Dionysos. And it is exactly these metamorphoses that are so impressive and lively. The same applies to the thorax aperture, the fleshy parts, we still sense the original edges of the cube forming the geometric thorax, but everything has been vitally changed. This is definitely one of the best European sculptures of genius, demonstrating the metamorphosis of stone into flesh and the cube into the human form. In this figure we can see how, by means of the geometric images of our visual cortex, nature is constructed anew; how it thus appears even more generally applicable, because it is created anew from our cerebral images, in the same way that Athene sprang from the head of Zeus.

If we compare the sculpture by Phidias to the photograph of a nude in a similar posture, the first thing we notice is that the nude has three creases across his stomach, caused by compression when the thorax is raised. Nevertheless, at the same time thorax and abdomen are more strongly interlaced as one torso, while in the sculpture by Phidias, although the creases are omitted, which is anti-naturalistic and actually illogical, the cube of the thorax is angled and set off more clearly against the cubes of the abdomen and pelvis. Through this much more cubic, geometric depiction of the torso the artist not only achieves a much higher degree of three-dimensionality, but we see far more clearly how heavily the torso is leaning on the supporting arm. The three parts, thorax, abdomen and pelvis are more clearly divided off and in spite of, or even because of, their cubic depiction are much more mobile, lie more heavily in their corporeality and are a lot clearer than the much more unclear forms of the natural

nude. This comparison shows very clearly that, according to current ideas, this figure is not very naturalistic. It is very much a new invention of our brain, where geometric perceptual images have been applied to natural objects.
Just as we recognize the geometrically simplified whole form first of all when seeing, the sculptor structures forms in a geometrically simplified way at the start of his work. While completing his work, these forms are then distorted step by step by the basic processes and by observing nature. Because of this, essential parts of the initial geometric structures are preserved up until the end.

He thus starts in a similar way to Wotruba, who has made this his principle style as seen in his two nudes (452, 453). They are only made of cubes, which have not been distorted by the basic processes and thus transformed into expanding, vital shapes. We can see how this extreme has led to flaccid, hanging shapes and a composition with weak expression. This is because unaltered, or almost unaltered, geometric forms are in themselves expressionless and meaningless.

There are, of course, completely different methods. Michelangelo's "Night", who is lying in a more curled-up posture, has more than natural stomach bulges which, together with her breasts, form an inward directed, three-dimensional plastic rosette (99). Michelangelo certainly created his figures in extremely distorted geometric forms from the start too, but they were nevertheless simpler and more Euclidean than after their completion, as many of Michelangelo's sculptures preserved at all stages of work show.

On the basis of a fairly small number of Euclidean geometric forms of our cerebral perceptual images, we are able to interpret an infinite multitude of natural forms and the artist can continually invent new forms. We can differentiate between whole schools of art by seeing which perceptual images they prefer as inspiration for their inventions.

To sum up: I would say that even after all the studying of nature that doubtless preceded the carving of these three sculptures (446, 447, 450), the sculptors looked anew at nature, and constructed anew, through the filter of their geometric perceptual concepts. Whether simple, as in my students' drawings, or complicated, as in the case of the classic and early Hellenistic Greeks, perception and its reproduction is a creative act, which, despite all observation of nature, reality invents anew through the metamorphoses of geometry.
I have no doubt at all that our brain tends to see reality more geometrically and more Euclidean than it is, perhaps because in the very act of visual perception we compare reality with the geometrically simplified contents of our visual memory, and, in so doing, alter it – as we definitely do if it is reproduced by the short or long-term memory.

It is interesting to look at the transformation of Greek sculpture by Roman copies from this point of view. We cannot fail to notice that although the Romans went to great pains to imitate the vitality of Greek sculpture they actually produced much more geometrically schematized work.

Let us consider the best of the surviving Roman copies of the originally Greek discus thrower (458). At first sight it appears to be close to nature; for example, the veins in the upper arm. In fact it is probably massively geometrically simplified, even compared with the Greek original – even in the sense of Hoffman's "orbits". We have placed a photograph of a nude (459) beside the detail of the discus thrower in which the course of the serratus muscle, starting from the lower tip of the shoulder blade and attached to the ribs at the side, and the ribs can be seen. Firstly, serratus and ribs are variously spaced and secondly, the serratus finishes diagonally above the ribs. In the Roman copy all the divisions of the ribs and serratus are identically wide and totally parallel. In the Greek bronze of Riace, one of the two figures of young men recently rescued from the sea, they are better differentiated (461). Serratus and ribs are different in direction and width. In the statue of the discus thrower we find the same schematization in the construction of the face and rib-vaulting, which completely separates the thorax from the abdomen in contrast to the Greek original, as we see in the ribs and serratus. For further comparison, let us take two Roman copies of the lost Greek originals of Polyklet's Doryphoros and of Diadumenos (460). We need not go into details, but here, too, we note the schematized geometrization, compared with Greek sculpture. Roman copies appear more sterile, colder and less vital, thus containing less visual information.

The degree to which original Greek art was, in the course of the centuries, retro-metamorphosed by the Romans into ever more schematized geometry can be shown by the example of the statue of a Roman consul in the Conservator Palace, second half of the fourth century (462). The original gesture of victory from the imperator statues has been transmuted into two arms bent at identical right angles, one pointing up, the other, down. The folds of the toga are all equally wide and deep and only grooved. There are almost no more rotated surfaces, apart from the folds under the raised arm. Former knowledge about the movements of a counterpoised human body has again given way to a rigid cubic block, with the knee of the non-supporting leg poking somewhat disconnectedly through the chiton. The metamorphoses of geometry correspond so little with the depicted object that the effect of the whole figure is sterile, not vital. Even if the relief on the tomb in Palmyra (463) is not one of the most successful works of its time, it is all the better for demonstrating schematism. Typical of the geometric bust schema are: the too long nose; the too short mouth/nose area; the identically wide and identically deeply carved rigid eyes; the jointless and identically wide fingers; the arms bent at right angles.

In the pre-geometric and geometric periods of archaic and early classic times, the various stages of the distortion of geometric forms into sculpture remained touching and vital because, step by step, they added something new to the old schema and distorted it by means of newly discovered possibilities, but late Roman sculpture seems sterile and lifeless, because, through lack of understanding, the mass of transmitted detail has been changed back into geometry; frozen, as it were.

In the following centuries of the migration of nations all the remaining Greek sculptural knowledge was destroyed and extinguished, but at the turn of the millennium there was a new beginning, for instance in the portal of St Zeno (twelfth century) in Verona (464, 465). On bases organized in a screen-like manner, from spiral derivations of Hoffman's "orbits", we see almost rectangular cubic figures. Their arms and legs are bent almost exclusively at right angles. With their parallel schematic lines suggesting pleats, and patterns scratched on the borders of their robes, these figures are truly the sort of new beginning we have described so often as happening in places all over the world. They have as good as nothing to do with the Greek figures of antiquity. This was the beginning of the development, in the second millennium, of the new European art, via Romanesque, early Gothic, Renaissance and finally arriving at Baroque and Rococo.

The geometrization of the St Zeno figures and their details demonstrates a similar stage of development as the famous bronze reliefs from Benin, Nigeria, sixteenth century. This latter development was probably brought to an abrupt end by the commencement of the Christian occident's slave trade with America.

But even the completely distorted Baroque and Rococo sculptures still show that they originated in our cerebral geometric concepts. Not only did the ornament which gave its name to the Rococo, the rocaille, develop from a circle via a C to an unevenly curved, and thus directed, C-flourish but even the sculptures are still based on geometric – though much changed – images (470, 471). I would like here to refer to one of Feuchtmeyer's famous Birnau putti (470). Even a quick glance shows that not only surrounding material and hair but also the entire body, legs, trunk and arms have been developed from the continually rediversified distortion of three-dimensional spirals. The entire composition of the "Ascension of the Virgin" in the monastery church of Rohr by the Asam brothers is also based on this geometric form. It was one of the basic geometric images of the dynamically mobile spaces and sculptures of the Rococo. I have already said that the four "orbits" of the Lie group and Hoffman are certainly only a part of our visual cortex's geometric inventions. We compare our environment with them and draw our conclusions and information from the differences between the object we are looking at and these concepts.

That is just one part of it. Actually we recognize from the differences to our geometric perceptual images the processes of growing and the fate and character of the objects we come across. However, when they are reproduced as works of art these objects are not only humanized in our sight by the more balanced and emphasized relation between natural form and geometrization – i.e. through the re-creation of reality – but over and above this seem to receive their universal meaning particularly from the metamorphoses of geometry (450).

The landscapes on which Cézanne modelled his paintings may have appeared fortuitous and more or less everyday, but in his paintings, through the new construction, through the reconstruction, of this reality from his brain's geometric concepts, they gained the universal validity we admire (472). "Street in Chantilly", 1888, consists of a composition of cubic or rectangular forms: the sections of the path; the shadows, the tree trunks; the walls of the house; then the sloping, pointed roof formations and as a third factor the unevenly vaulted foliage – green, violet, blue and ochre. Basically the whole diverse landscape is composed with dynamic brush-strokes from three kinds of geometric form – rectangles or cubes, prisms, segments of circles – more, or less, distorted by the basic processes. Despite these alterations, some of which are massive, and the newly invented forms within, for instance, the vegetation, the landscape has lost none of its variety, but, because of its re-creation through our mind, has changed from a more or less fortuitous section of a view to a highly unified painting of universal validity.

In the same way the "House in Provence", 1885/86 (473), is an example of what is basically a common southern landscape, which, through distorted rectangles and cubes and dislocated unevenly curved parallels has made of such various motifs as rocks, vegetation, houses, paths and fields, a unified, universally valid bare southern landscape. We are impressed not by unique motifs, but by the powerful synthesis of human perception in this admirable painting.

We have arrived at late Impressionism. The unification of nature with our cerebral geometric images was virtually the programme of its practitioners. We have already analyzed a drawing by van Gogh. Others I could mention include Seurat, Delaunay and the young Picasso, not forgetting German Expressionism.

The late Impressionists' interest in emphasizing more strongly the geometric concepts in their pictures certainly led to the discovery of African sculpture. Except for a short period between the thirteenth and sixteenth centuries in Nigeria, African sculptors, probably because of the fetishist, religious, cultic character of their work, were less interested in individual reproductions of nature than in depictions of gods and demons, in fetishes, in the unification of separate human individuals to a cultic congregation. That is why in Africa these products were not thought of as art; why in the thousand or so African languages there is no word for "art".

At the beginning of the twentieth century, when, because of the preconditions described above, Europeans discovered these products, they judged them to be art and used them, which led, in some cases, to outstanding work, such as Zadkine's "The Destroyed City" in Rotterdam (474). But frequently it inspired a more aesthetic-eclectic collage of European and African elements. One famous example of this is Picasso's "Demoiselles d'Avignon" in which incomprehensible fetish-like aspects are combined with European counterpoise movements. It was an attempt to invent a new kind of art (475).

Of course, Zadkine's "The Destroyed City" was not only a new invention, in form quite free compared to the African models; quite apart from this the sculpture was concerned with a non-individual, demonic-apocalyptic event of universal application. Greek-European individualism was perhaps unsuited to its depiction. But this is exactly why this combination of African and European carving and painting – some German Expression-ists also produced African sculpture which had little to do with their contemporary painting – was limited to one generation. It was not a new beginning, but a long-abandoned experiment, which had produced great achievements, to actually reconcile irreconcilables. It did not develop into a new perception of art in European and dependent cultural circles.

At most it was responsible in the former century for a flatter and more cubic approach in European figurative sculpture or one more strongly oriented towards basic geometric bodies, from Lehmbruck (476), Barlach (477) and Marcks (478), Wimmer (479), Seitz (480), Grzimek (481), Hrdlicka (482) and Weber (483). But that would probably have happened even without the trip to Africa – through Cézanne or the German Hildebrand and Marées alone. By the way, even if none of these painters produced his work with the intention of being modern, of creating a sensation in order to be noticed, it is clearly the style of the twentieth century. This kind of sculpture cannot be confused with work from any other century on any continent. It just so happens that hardly anyone has noticed yet it is called traditional.

This happened because of that trip to Africa, which led to a completely false judgement of European art, especially after the Second World War. In Germany, in the fifties and sixties, High Renaissance and Baroque art were condemned as inartistic naturalism. Let me remind you of Willi Baumeister's book "Das Unbekannte in der Kunst" (The Unknown in Art), 1947, in which he just about permitted Dürer and others still to count despite their "naturalism", because they were such great artists that even the "Unknown" had then flowed into their work. No, these works never were naturalistic; they were close to nature; they were nature newly originating from our cerebral perceptual images; they were a new idea of reality, as represented in the incomparable Greek symbol for the birth of an idea: "Athene springing from the head of Zeus". Through the clumsy comparison with African art and the simultaneously beginning abstract art, this was overlooked.

So it is not surprising that most of the art of the second half of the preceding century swung backwards and forwards in a directionless manner between the reproduction of photographically produced pictures, casts from nature and the alleged new or re-discovery of square, rectangle, circle, triangle and parallels.

Let us start with the photograph, which actually does use technical means to reproduce reality as it appears on our retina and not changed, not newly created, newly invented by our brain. Some of the "photographic painters", such as Richter (493), tried to hide this by painting the photographic reproduction out of focus, in order to create an illusion of Impressionism, reminiscent, from a distance, of Monet – a mixture of mindless technique and shameless eclecticism. However, these were just technical tricks, and not a new invention from the spirit of our visual cortical areas.

It was the twentieth century, inspired by some nineteenth-century casts from nature, which invented real naturalism, a reproduction by means of photographs or casts, showing nature in new material but not in a new spirit. Creativity was excluded by technical processes. There was no longer any call for altering our environment by means of our perception and thereby humanizing it (494 – 496). Some, like Segal (494), tried, instead of new formal and intellectual invention, to attain a degree of "alienation" by means of unaltered, unpainted plaster casts. As a general rule the word "alienation" began to replace the invention of form and colour. Other people painted their casts with naturalistic colours and used real hair, headscarves and real jackets; they dressed their dummies, they filled the shopping trolley with real goods (495). The end of all art, but also of all conjuration of demons, gods or devils. It was mental dullness transformed by technology; and thus, as such, a kind of allegory of the twentieth century.

This Capitalist Realism – Richter invented this term, his only real invention – was actually the same as the Soviet Union's Socialist Realism. The former declared consumer goods to be art and the latter concealed inhuman and criminal ideas with pretended ideals and apparently artistic means. Since form and content cannot be separated – as we determined when analyzing fruit and plants – both methods were expressions of mendacity and mental dullness.

On the other hand, people were claiming to have invented the square, the circle, parallels, the right-angled cross and the screen anew (484 – 491) – only this time using ruler and compass rather than drawing freehand as in previous artistic beginnings, and claiming to be able to make cosmic pronunciations about our existence by these means. While, as we have seen, in the case of the earlier beginnings the alternation between the geometric forms for the purposes of decoration brought about a certain variety, these above-mentioned Euclidean forms now frequently appeared one by one (484 – 486). But there is no getting away from the fact that the basic Euclidean forms are perceptual images of our visual cortex. We measure and recognize the millions of processes and phenomena around us against them and because of this they have no independent meaning and expression.

Like light, the basic Euclidean geometric forms, the schematism of simplicity, are the precondition for our visual perception. A stroke of genius by nature, which probably existed in the initial stages millions of years before the creation of mankind on earth. To hold up these geometric forms as a new artistic invention, in view of the fact that we can look back on at least some 30,000 to 40,000 years of artistic development – and know children's drawings – can only be explained as ignorance of how our visual recognition works. Nor must we forget that these simple geometric forms have been combined with American advertising for consumer goods. Everyone knows Coca-Cola bottles (492), canned meat, emblazoned T-shirts, sex symbols. They have begun their triumphal progress around the world helped by the media. They have caused the degeneration of the perceptual area in our brain in a manner reminiscent of the fate of Kaspar Hauser (a famous nineteenth-century case of a child who grew up in complete isolation). In more than forty years of teaching in a university I have witnessed a steady decline of perceptual abilities in our young people.

The exhibition "Ornament und Abstraktion"[48] in Basle from 10 June – 2 October 2001 showed certain similarities of such "modern" geometric forms with early ornamentation. There is surely no doubt about the direct dependence – art from all over the world has, after all, been available to us in books for more than a century and we know that many artists have used such publications – but nevertheless we should not underestimate mentally based repetition. Malevich certainly did not need any prehistoric stimuli to rediscover the circle or the square. They belong to our brain's perceptual concepts.

It is no wonder that a time in which the basic geometric forms which the infant recognizes in its first year and reproduces in its fourth and fifth year, which now hang in exhibitions as works of art, should even officially regard as art, pictures analogous to infants' first scribbles; i.e. up-and-down movement, or as the American John Matthews calls it, the push and pull (cf. 498, 501 and 16). The pictures consisting of curvy lines, loops, multiple loops, spirals (Lotte, c. 2 years old) are strongly reminiscent of corresponding pictures by André Masson, Twombly and Pollock. The way Lazslo, 15 months, ranged across a surface (17) is reminiscent of the works of the Dutch groups, Appel (499). It is scarcely necessary to comment on the amazing resemblance between the vertical strokes hitting the surface or the floor, with subsequent loops (21 and 20), by Lotte (24 months) and Pollock, who also used to lie on the floor to paint.

In the meantime we have learnt that this Abstract Expressionism – the name given by the Americans to these paintings making use of large formats and oils to repeat in a rather more disciplined way the pictures children produce in their second or third year – was deliberately encouraged by the American CIA as a counterweight to the Soviet Union's Socialist Realism. This probably accounts for the rapid triumphal progress of these pictures around the western world – another kind of official state art! Pollock and de Kooning remain its most famous practitioners. At least in the case of Pollock, correspondence with children's drawings in the second half of their second and first half of their third year matches in every detail.

I am not saying that such traces of movement are completely expressionless; after all, they prepare the development of the geometric forms and, to that extent, are pre-forms, metamorphoses of these forms before the development of their pure shape. But it is no coincidence that Abstract Expressionism and the basic geometric forms developed simultaneously as art forms until the sixties.

Perhaps the macaroni lines of ice-age man, 35,000 to 30,000 BC, looked similar. The difference is that then, and at the beginning of the historical period, in the fifth and fourth centuries BC, these were beginnings, which developed further, while what we have now are final products, which have arisen simultaneously with the art which developed at the end of the nineteenth and beginning of the twentieth century from late Impressionism and a new cubic view of the figure (476 – 483). These cannot be confused with the artistic products of any other age or continent.

But let us return to visual perception and to consciousness, to the metamorphoses of geometry. Of course, after developing this theory I have no wish to go on and give a detailed definition of our human consciousness, which certainly differs profoundly from animal consciousness, even that of the primates. In my opinion it is dependent on many things: our attentiveness, our corporeal awareness, our sexuality and much more, but I am sure that the memory is one of the main components of our consciousness, both the identification memory and the reproduction memory, the short-term memory and the long-term memory. These differences are to be found in the entire memory area. I recognize words without their occurring to me in view of an associated object. I remember stories when I hear part of them again.

[48] "Ornament und Abstraktion". Foundation Beyeler, Basle

It can even be a particular smell that reminds me of a particular event.

But the difference between identification memory and reproduction memory is most obvious in the area of visual perception. The richer our long – term memories, the more developed our consciousness. The more general our memories, the less personal our consciousness. The development of the reproduction memory is thus also a question of the maturing of distinct personalities with real, personal, individual consciousness. Perhaps we should support the perceptual training of young people, the maturing of their personal visual memory more strongly and above all with better methods. In this way the self organization[49] of our minds would take place more harmoniously.

[49] Wolf Singer, "Zur Selbstorganisation kognitiver Strukturen", in "Gehirn und Bewußtsein", op. cit.

SUMMING-UP

In my examination of these questions I was primarily interested in how environmental phenomena, natural and man-made forms pass on information to us, how their expression comes about. In the course of our deliberations we came across the function of the geometric Euclidean forms, which are few and simple when compared to complicated natural phenomena.

The gestalt psychologists were the first to state that basic geometric forms played a decisive role in visual perception and recognition. They realized that in the first place we recognize environmental phenomena as a whole, as comparatively simple geometric forms such as, for instance, the human figures of fourth-millennium ancient Egypt (353). Then, according to the gestalt psychologists, we go on to recognize these simple total forms as consisting of separate, differentiated shapes, once again, of course, geometrically simplified. In their opinion we do not recognize and understand our environment until we have found a correspondence with these simple forms; when, for instance, as Arnheim writes, we have comprehended the tangle of trees, bushes and hills as a combination of rectangles, triangles, circles, parallel directions etc.

Following a completely different path, the mathematicians, using the Lie theory of continual transformation groups, realized that our visual cortex produces simple geometric forms: circles, radials, parallels etc., and then makes them correspond with environmental phenomena, using a method none of these mathematicians can clearly describe.

These were fertile beginnings, which, although starting from completely different points, arrived at the result that certain geometric forms play a very important part in our visual perception and that, somehow or other, we make environmental phenomena connect with them.

However, this path does not lead to the answering of my question on the information contained in forms, because the relatively small number of basic geometric forms is insufficient to explain nature's infinite variety and expression. It is also not able to explain comprehensively the gestalt psychologists' "first law of perception" on the spatial effect of complicated organic forms.

Calling on a very large number of examples I have attempted to prove that our visual judgement depends not on the correspondence with a smaller or larger number of geometric forms but on the difference to them. All that the simple geometric forms – circle, square, rectangle, triangle, parallels, radials, regular patterns etc. – tell us is that they exist, that we are dealing with a circle, a rectangle or parallels. Over and above this they cannot pass on any information; they are too general to express anything, to have a meaning of their own, let alone a dynamic.

But because we recognize the infinite variety of environmental phenomena, and the processes from which they have developed, by comparing them to the basic geometric forms and registering the differences between them, insofar as they can be judged visually, these basic geometric forms are themselves completely neutral, without expression and dynamic. It is the difference to them that we register as having occurred through movement, dynamic, some processes or other, and which shows us the expression of a natural, lively form, or one that we ourselves have made. We compare unevenly and evenly spaced radials; identical and different sizes; varying and identical densities; similar and dissimilar curvatures; rotated and parallel surfaces – and register the differences as movement, origin and individual expression of the natural forms. Euclidean geometric forms can be distorted in an infinite number of ways. Step by step, via schemata that grow ever closer to nature, this process can lead to more and more complexity and so our brains' geometric or geometry-like inventions become our visual criteria for surveying natural forms. We comprehend the differences between them as movement, as life, and thereby as their individual expression.

This step by step construction of the geometric perceptual concepts, via their combination with simple schemata as done in childhood, then step by step to ever more complicated schemata, with which we compare every natural phenomena corresponding to them, is no doubt a "bottom-up" and "top-down" process. This very process leads us to our judgement about the expression of natural phenomena or art. Experiments have to verify whether it is the same one that the neurophysiologists have proposed.

Because there are an infinite number of ways of diverging from the relatively small number of simple basic geometric forms – as we exemplified by means of trees, clouds, leaves, human and animal figures etc. – a comparatively small store of Euclidean forms – a cerebral invention of ours, almost never to be met with in nature – is indeed sufficient to recognize the processes which have led us to the visual phenomena around us. We perceive through comparison. I passed on my initial discoveries regarding this matter, the conclusions I had come to through my work as a sculptor, in my book "Gestalt, Bewegung, Farbe", 1975. We can recognize growth and decay, threats or aspirations, as expressions of natural phenomena. Our brain's basic geometric forms are the standard against which we measure our environment.

We used the development of infants' drawing to prove this point. We were able to show how the basic geometric forms develop from the basic processes: ups and downs, parallel displacement, expansion, contraction and rotation, and how the infant uses the forms to construct our environmental phenomena and to store them in its memory, so they can be reproduced. This accounts for the development of the visual reproduction memory, a part of our declaration memory. The figure schemata made from these geometric forms are the content of this reproduction memory. It usually continues developing until the child reaches its twelfth year. Nowadays, in most people, it fails to progress beyond this level. To counterpoint this we then described the degeneration of the reproduction memory and eventually of the identification memory of people suffering from agnosia and Alzheimer's disease.

The next step is then to use the basic processes to deform the schemata constructed from these basic forms. These then fill our reproduction memory with more lively schemata, which are closer to nature and with which we compare our environment. This enables us to achieve an ever more differentiated visual perception and recognition. The basic processes produce the Euclidean forms, from which the schemata of our reproduction memory are assembled in the initial stage. They are then distorted by the same basic processes from which they arose.

While our reproduction memory continues to improve for years, if not decades, our visual cortex, in its first two years, develops the identification memory, which works only when there is visual confrontation. We are not able, without visible confrontation, to describe precisely our fellow beings from memory, let alone draw them, but if we meet them we recognize them immediately. Indeed, we recognize them so precisely that we believe we can tell how they are feeling, in mind and body. The identikits of the police, which are based on descriptions, could be mentioned here. In reality these are produced by combining the reproduction and identification memory: based on a description, the artist draws pictures that resemble the person seen until the witness believes that he has recognized that person. It is quite obvious that this identification memory constitutes itself in the first eighteen months of life. It starts by recognizing a chessboard pattern or a square and has reached maturity when it can recognize people or places it has not seen for months.

The beginning of the scribbles which finally turn into the basic geometric forms starts around the time when the identification memory matures. That is the beginning of the reproduction memory. Its initial commencement cannot be really confirmed until the child has developed the basic geometric forms and begins to use them to reconstruct its environment – i.e. in its fourth and fifth year. This may well be the reason why hardly anyone has memories from early infancy. It seems that the largest part of the general declaration memory does not function for a long time without visual memories. In his autobiography "Jeunes années"[50] Julien Green wrote that his memory in particular recollects things he has seen more than things he has heard – apart from music. I do not, however, think that is something particular to him but rather that it applies to everyone.

The identification memory is to be found in every area. We recognize words passively which we would not use actively. The visual identification memory is probably dependent on the basic geometric forms, which are enriched by details but cannot be called upon except in the case of visual confrontation.

By showing examples of the early ornamental styles and the earliest figures from every part of

[50] Julien Green "Jeunes années". Paris, 1984

the world we proved that the basic geometric forms are the most important building blocks for our reproduction memory. Ornaments everywhere begin with the same geometric forms: the chessboard pattern; the circle; its radials; zigzag and parallel lines. These then, distorted by the basic processes, develop away from each other in the various parts of the world.

We make use of the same basic geometric forms when remembering people, animals and plants and when trying to reproduce them. The next step is for them, too, to be distorted by the basic processes and finally to be transformed by observation of human, animal and plant movement, though admittedly simplified and schematized. The great correspondence between the beginnings of all depiction of ornaments and figures all over the world, in every culture at different times, did not come about because of cultural export and import, but has a mental base, because all over the world the visual cortex produces the same basic geometric forms. They, like light, are the basis of every visual perception.

During the later course of history, when land and sea trade brought the whole world closer together, artistic culture quite clearly developed differently, even when there was mutual influence. I am thinking, for instance, of the great difference between Chinese, European, African and American culture in the middle of the fifteenth century. The great similarity between all cultural beginnings originates in the way the visual cortex functions and not in trade and traffic as Heyerdahl assumed.

Geometry's greatest triumphs have, of course, been celebrated in the field of architecture. However, we have not drawn our examples from architecture, as many rational reasons can be advanced for the use of geometric forms when dealing with such purely geometrical things as rectangular ground plans, vaulting and circular-shaped individual buildings. But at this point I should like to add that, in the course of the history of architecture, forms altered by the basic processes soon developed from the purely geometric buildings. The entasis of the Doric column with its unevenly curved shaft and the uneven curvature of the echinus express expansion and burden. There is no static or constructional reason for this form. If anything, this distortion of the basic geometric form made construction more complicated. But now everyone could see that the columns expressed natural tension through the comparison between the geometric cylinder and the unevenness of the curve of the entasis. As architecture developed further, in the Gothic, Baroque and Rococo periods of the western world, the purely perceptual based distortion of geometric forms by imagined basic processes gained ever more influence, until it was one of the leading factors in design, without any practical reason in statics. On the contrary, the statics frequently became even more difficult for this reason, as shown in the monastery church in Neresheim by Balthasar Neumann. It was left to the modern architects of the twentieth century to put a radical end to this organic, plastic development of expression and go back to using purely geometric forms – only more exactly and thus more undistored – as in the beginning of history. That is the reason for modern architecture's frequent sterility and lack of expression.

We perceive through comparison. We compare all natural forms with the geometric forms invented by our brain and this enables us to see the expression of growth and decay, aggression and peacefulness, joy and sadness. In a similar fashion we express ourselves artistically. In art the point of departure for work is sometimes these geometric forms and sometimes those. Distortions can be slighter or more dynamic. In art the metamorphoses of geometry appear purer, more consistent and more free of coincidence than in the natural world around us and precisely for this reason they are a clear indication – clearer than the results of many experiments – of how our visual cortex functions.

BIBLIOGRAPHY

Arnheim, Rudolf "Kunst und Sehen", Berlin 1964 ("Art and Visual Perception", Berkeley, Los Angeles, London 1954)

Attneave F., "Some Informational Aspects of Visual Perception", Psychological Review, 1954

Carter, Rita "Mapping the Mind", London 1998

Crick, Francis "The Astonishing Hypothesis: The Scientific Search for the Soul", London 1994

Dodwell, Peter C. "The Lie Transformation Group Model of Visual Perception", Ontario 1983

Dodwell, Peter C.; Wilkinson, France E.; Grünau, Michael W. von "Pattern Recognition in Kittens: Performance on Lie Patterns", Ontario 1983

Dodwell, Peter C. and others: "Handbook of Infant Perception From Perception to Cognition", Florida 1987

Drösler, Jan (lecture in Braunschweig)

Engel, A.K.; Fries, P.; Singer, W. "Dynamic Predictions: Oscillations and Synchrony in Top-Down Processing" in: Reviews, Nature Neuroscience, Volume 2, October 2001

Fantz, Robert 1960

Frisch, K. von "Die Tanzsprache und Orientierung der Bienen", Springer, Berlin Heidelberg New York 1965 ("The Dance Language and Orientation of Bees", Cambridge Mass. 1967)

Gibson, James J. "The Perception of the Visual World", Boston 1950

Golomb, Claire "The Child's Creation of a Pictorial World", Berkeley 1991

Green, Julien French original "Jeunes années", Paris 1984

Hawking, Stephen "The Illustrated A Brief History of Time", Bantam Books, New York 1996

Hölscher, Tonio "Klassische Archäologie - Grundwissen", Darmstadt 2002

Kebeck, G. "Wahrnehmung", Munich 1994

Kellog, R. "Analyzing Children's Art", Palo Alto CA: Mayfield, 1969

Koffka, Kurt "Principles of Gestalt Psychology", New York 1935

Köhler, Wolfgang "Intelligenzprüfungen an Menschenaffen", Berlin 1921

Korfmann, Manfred "Troia als Drehscheibe des Handels im 2. und 3. vorchristlichen Jahrtausend" in "Traum und Wirklichkeit in Troia", 2001

Maurer, Konrad; Maurer Ulrike; Horn, Tilde and Frölich, Lutz "Alzheimer und Kunst", Vernissage Carolus Horn

Maurer, D. and Barrera, M. "Infants' Perceptions of Natural and Distorted Arrangements of a Schematic Face", 1981, Child Development, 52, 196-202

Matthews, J. "Children Drawing: Are Young Children Really Scribbling?", 1984, Early Child Development and Care, 18, 1-39

McClung, Robert M. "Die Tarnung der Tiere", Wien 1978 ("How Animals Hide", Washington 1973)

Pöppel, Ernst "Eine neurophysiologische Definition des Zustandes Bewußt", in: Gehirn und Bewußtsein, Weinheim 1989

Sacks, Oliver "An Anthropologist on Mars: Seven Paradoxical Tales", New York 1995

Sacks, Oliver "The Man Who Mistook His Wife for a Hat", 1985

Singer, Wolf "Zur Selbstorganisation kognitiver Strukturen", in: Gehirn und Bewußtsein, Weinheim 1989

Weber, Jürgen "Gestalt, Bewegung, Farbe", 3rd edition, Braunschweig/Berlin 1975/1984

Weber, Jürgen "Abitur - Hochschulreife?", in: BDK-Mitteilungen, Fachzeitschrift des Bundes Deutscher Kunsterzieher e.V., 1/95, p. 25-27

Zeitgeist-Gesellschaft, Berlin, Catalogue "Afrika", Exhibition 1 March - 1 May 1996

INDEX

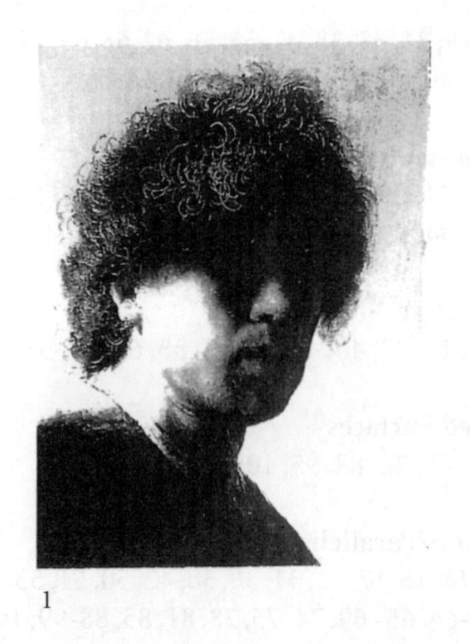

1

2

3

4

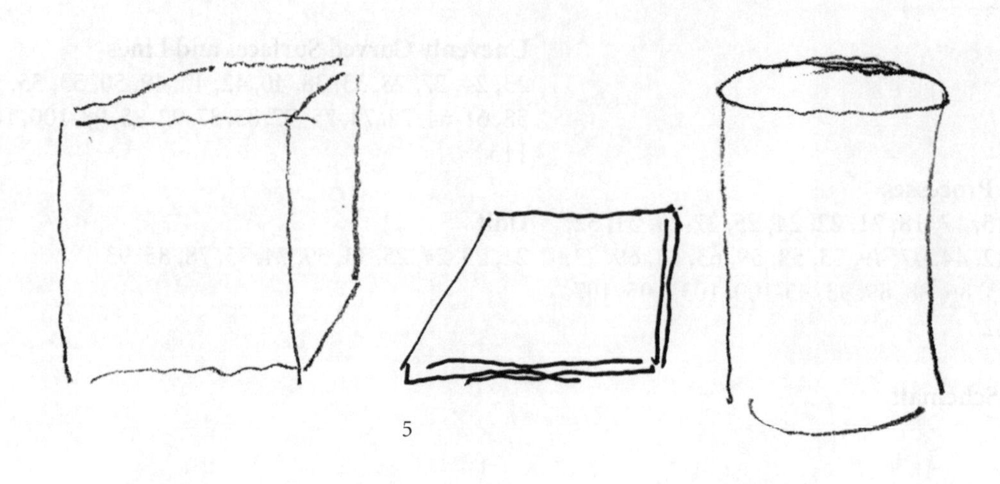

5

6

7

8

9

10

11

12

13

14

15

16

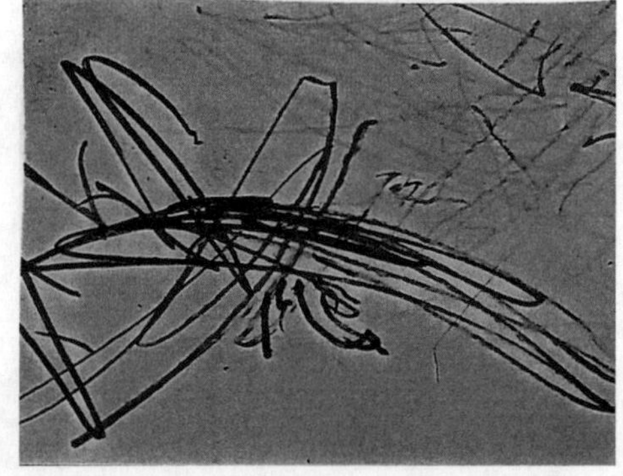

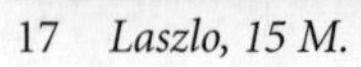

17 *Laszlo, 15 M.*

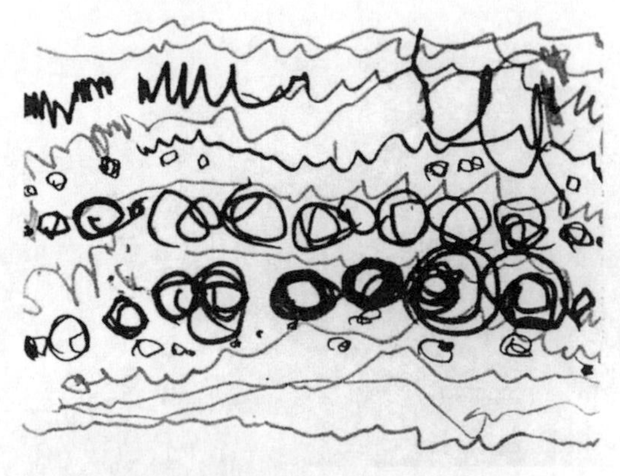

18 *Lotte, 2*

19 *Lotte, 2*

20 *Pollock*

21 *Lotte, 2*

22

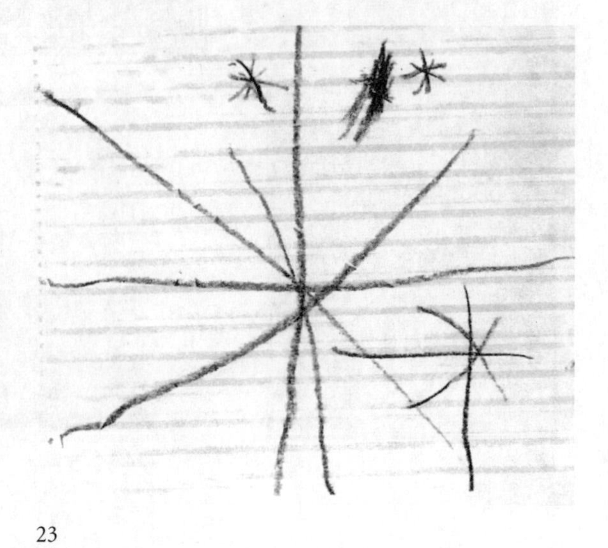

23

24 25 26 27 28 29 30 31

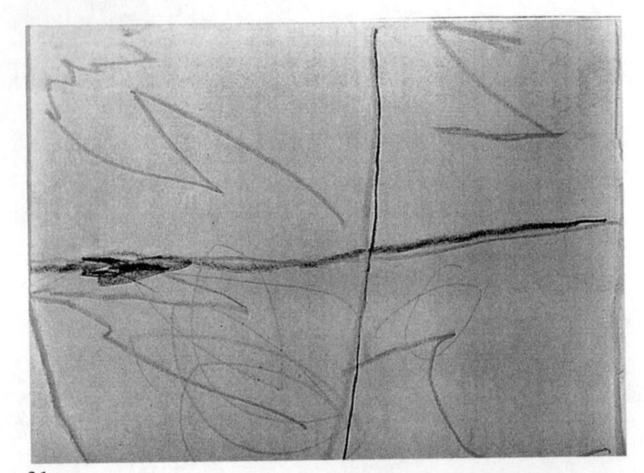

31 a

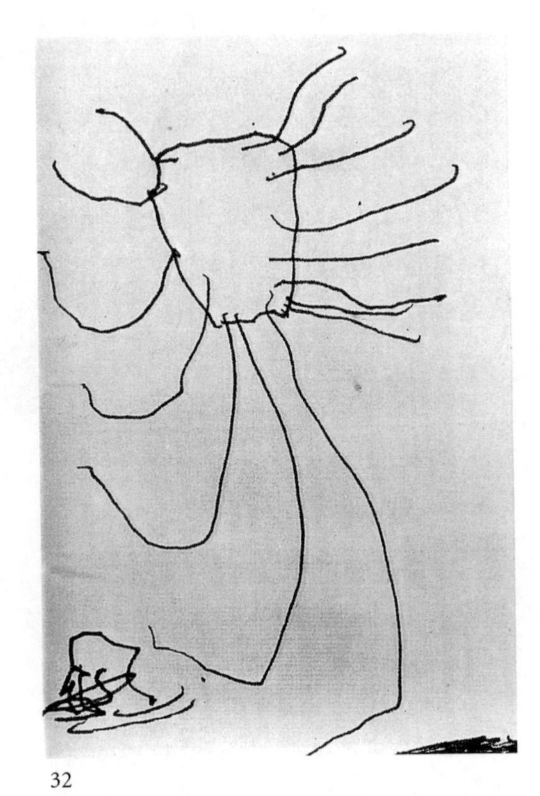

32

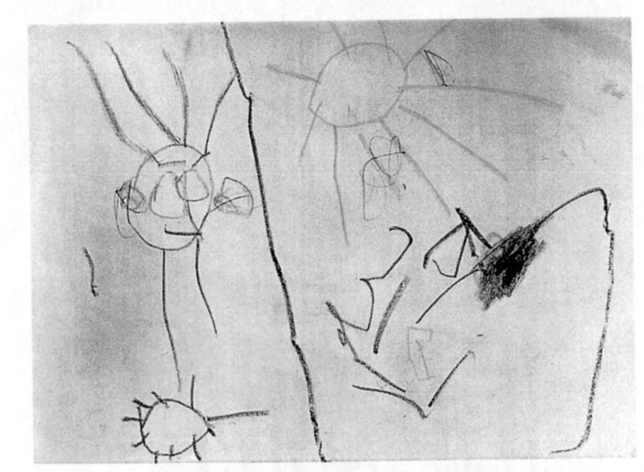

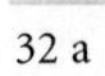

32 a

33

34

35

36

37

38 *Constantin, 5*

39

40

41 *Katja, 8*

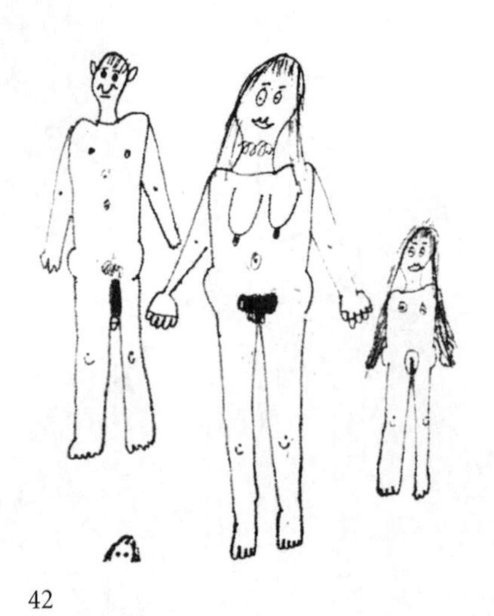

42

43

44

45 *Saskia, 9*

46

47

48

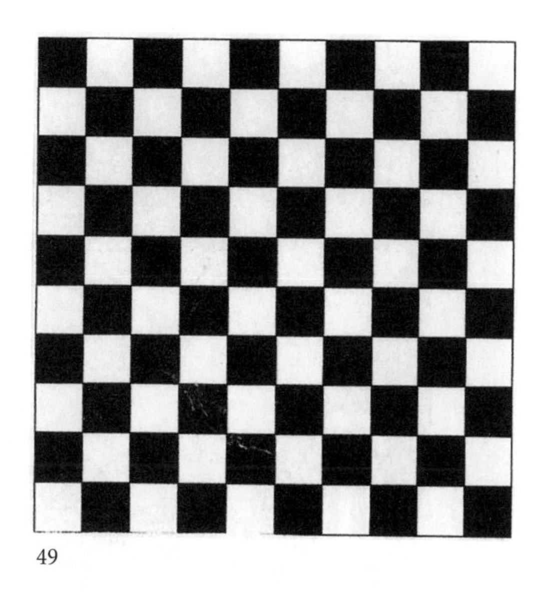

49

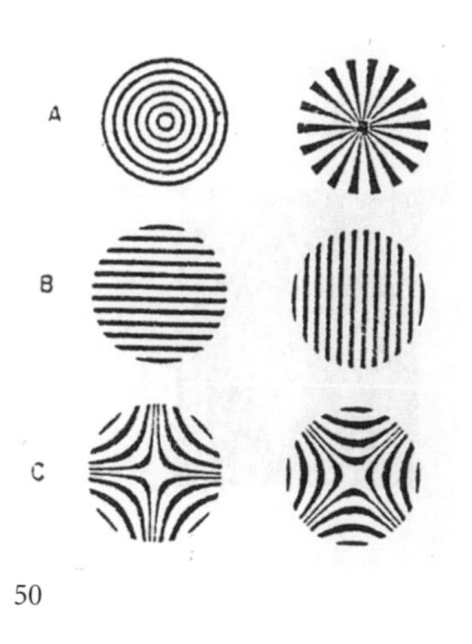

50

51

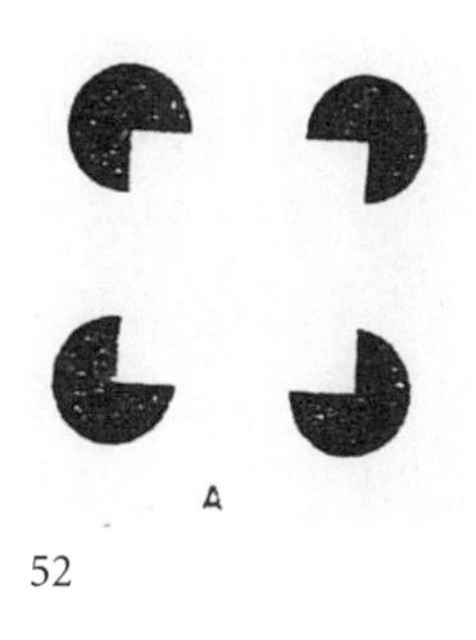

52

53

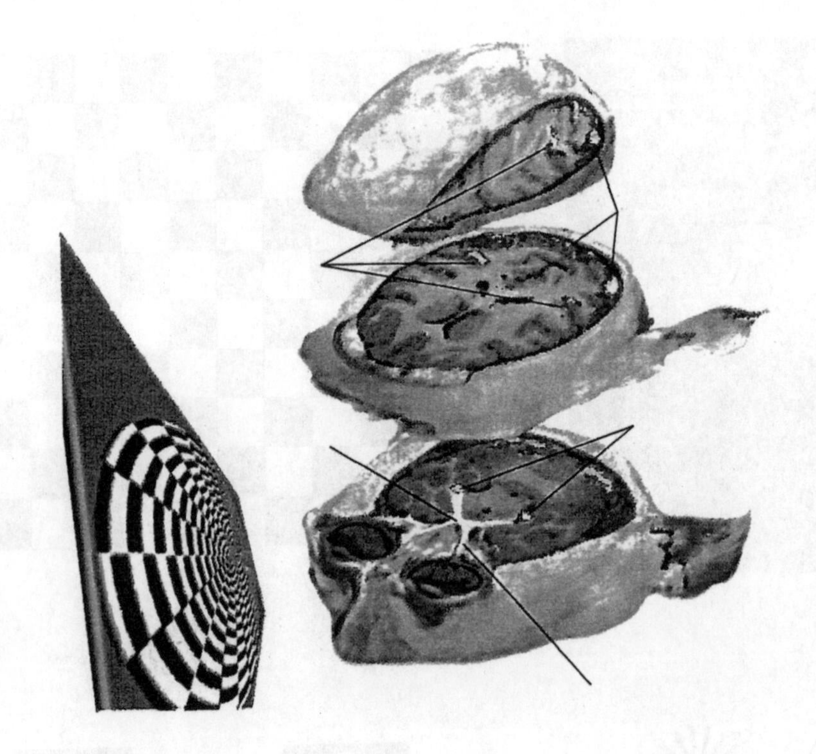

54

55

56

57

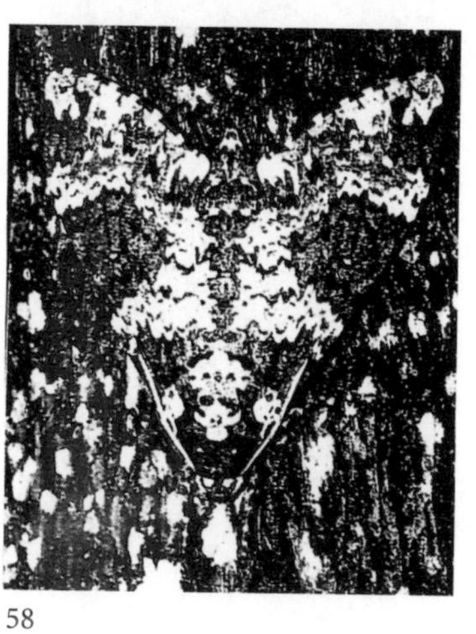

58

59

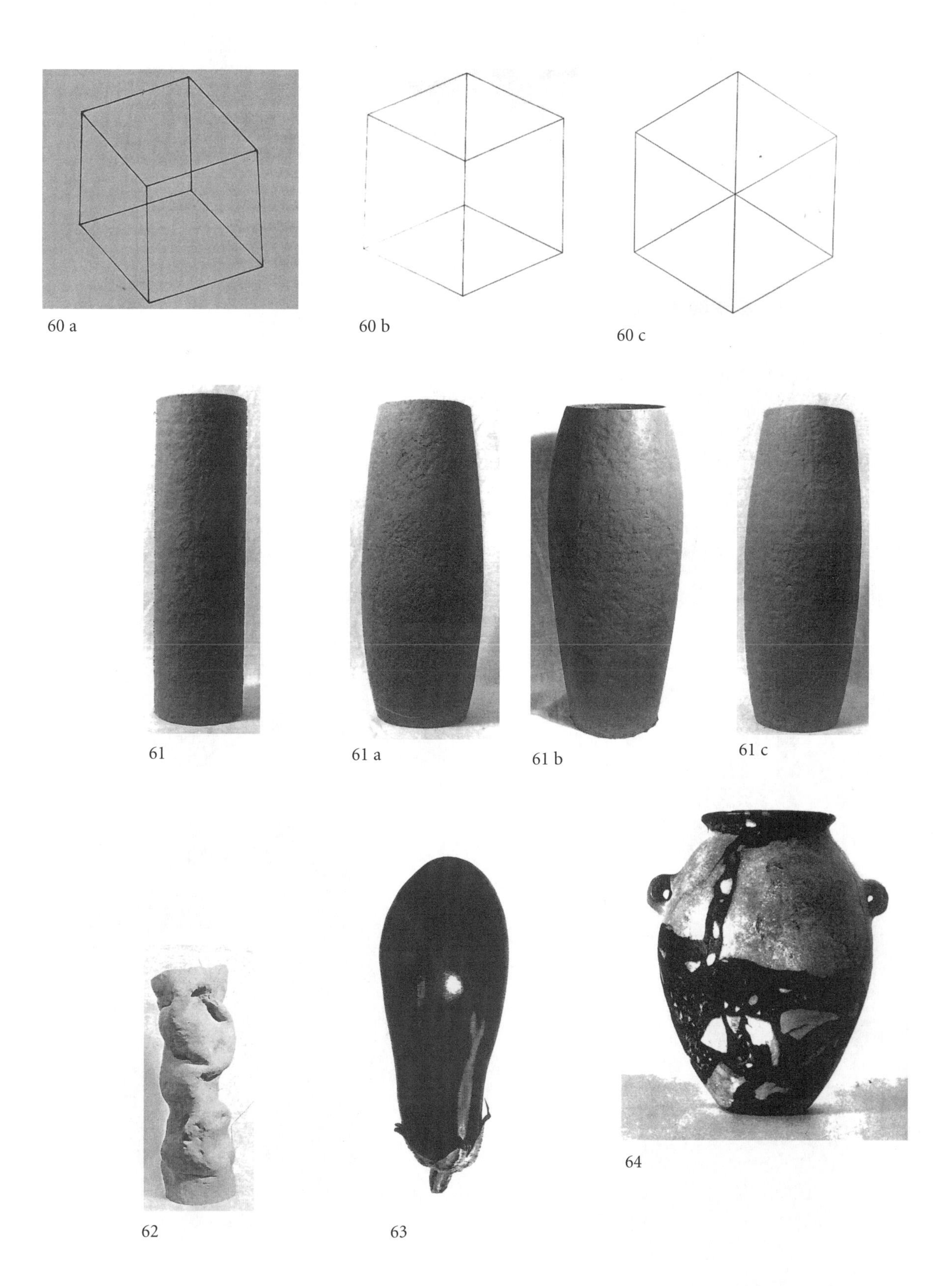

60 a
60 b
60 c

61
61 a
61 b
61 c

62
63
64

65

66

68

67

70

69

72

71

73

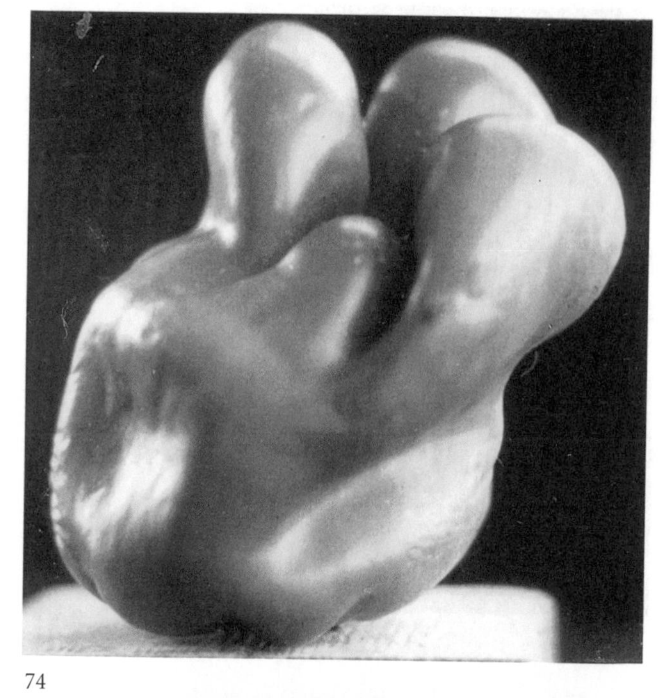

74

75

76

77

78

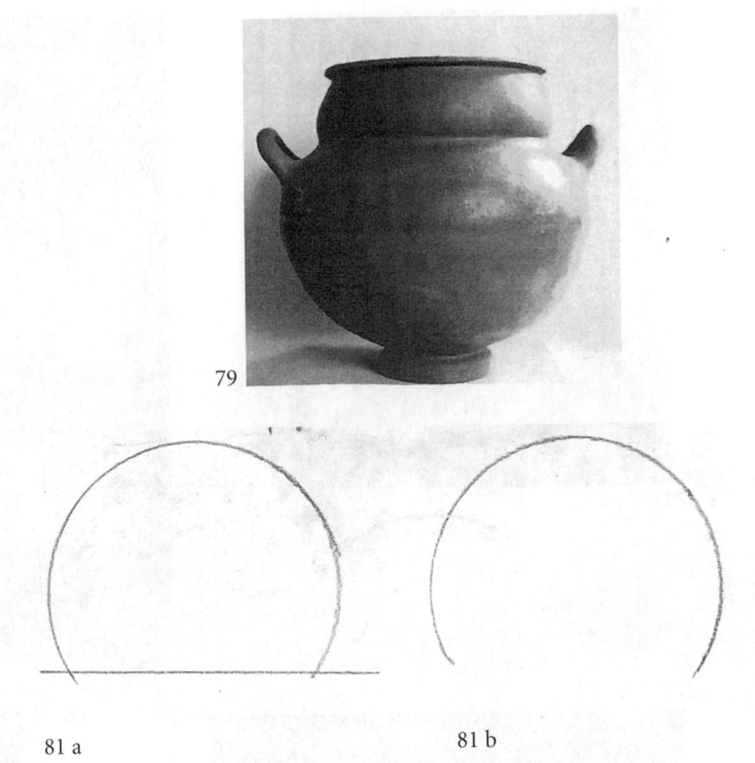

79

81 a

81 b

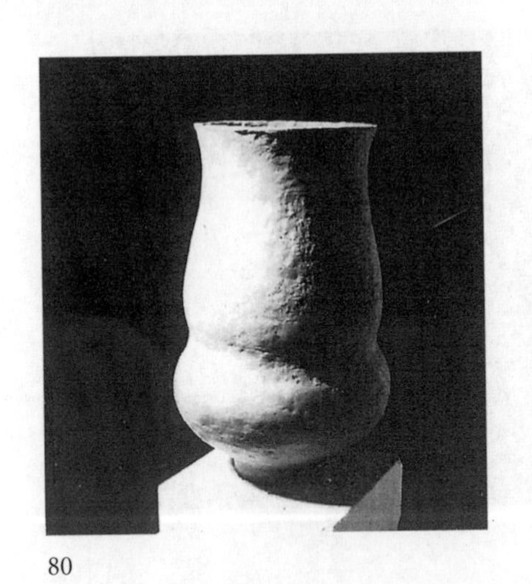

80

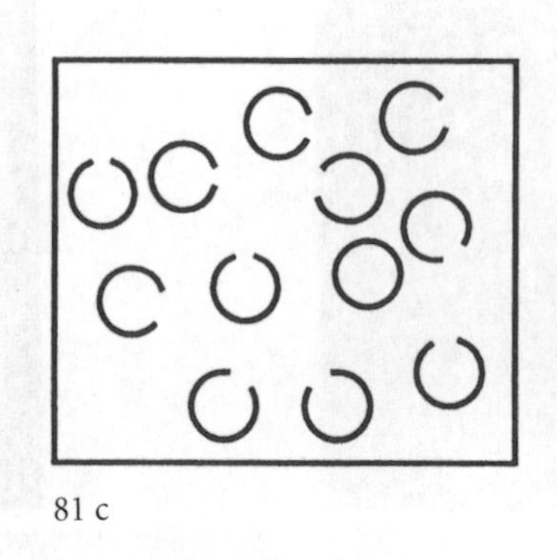

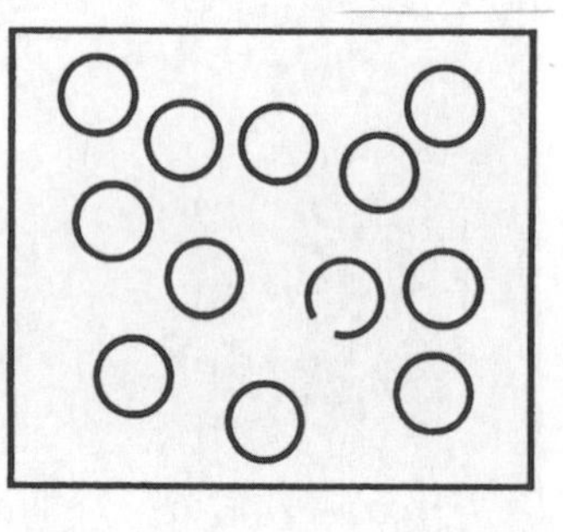

81 c

82

83

84 85 86 87 88 89 90 91 92 93

94

95

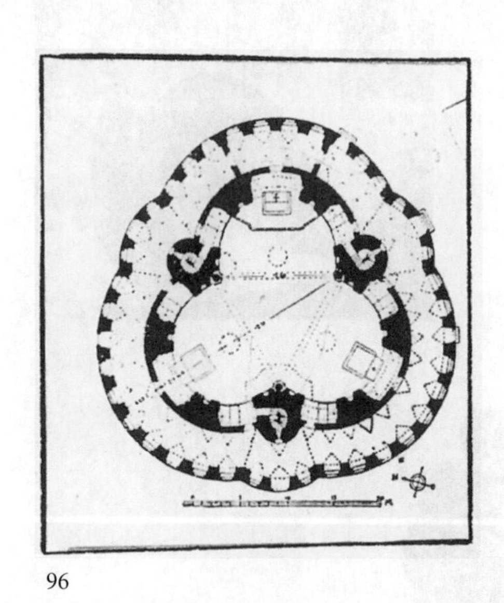

96

97

98

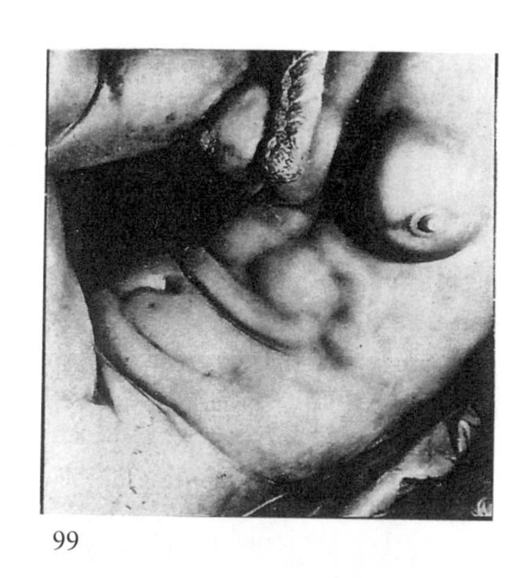

99

100

101

102

103

104

105

106

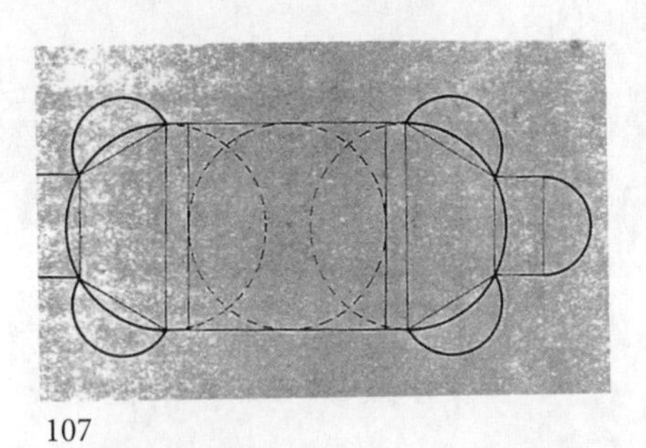

107

108

109

110

111

112

113

114

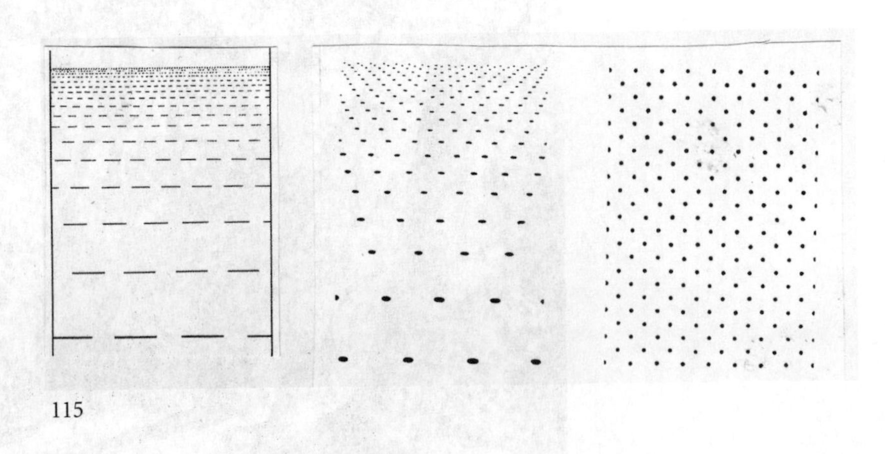
115

116

117

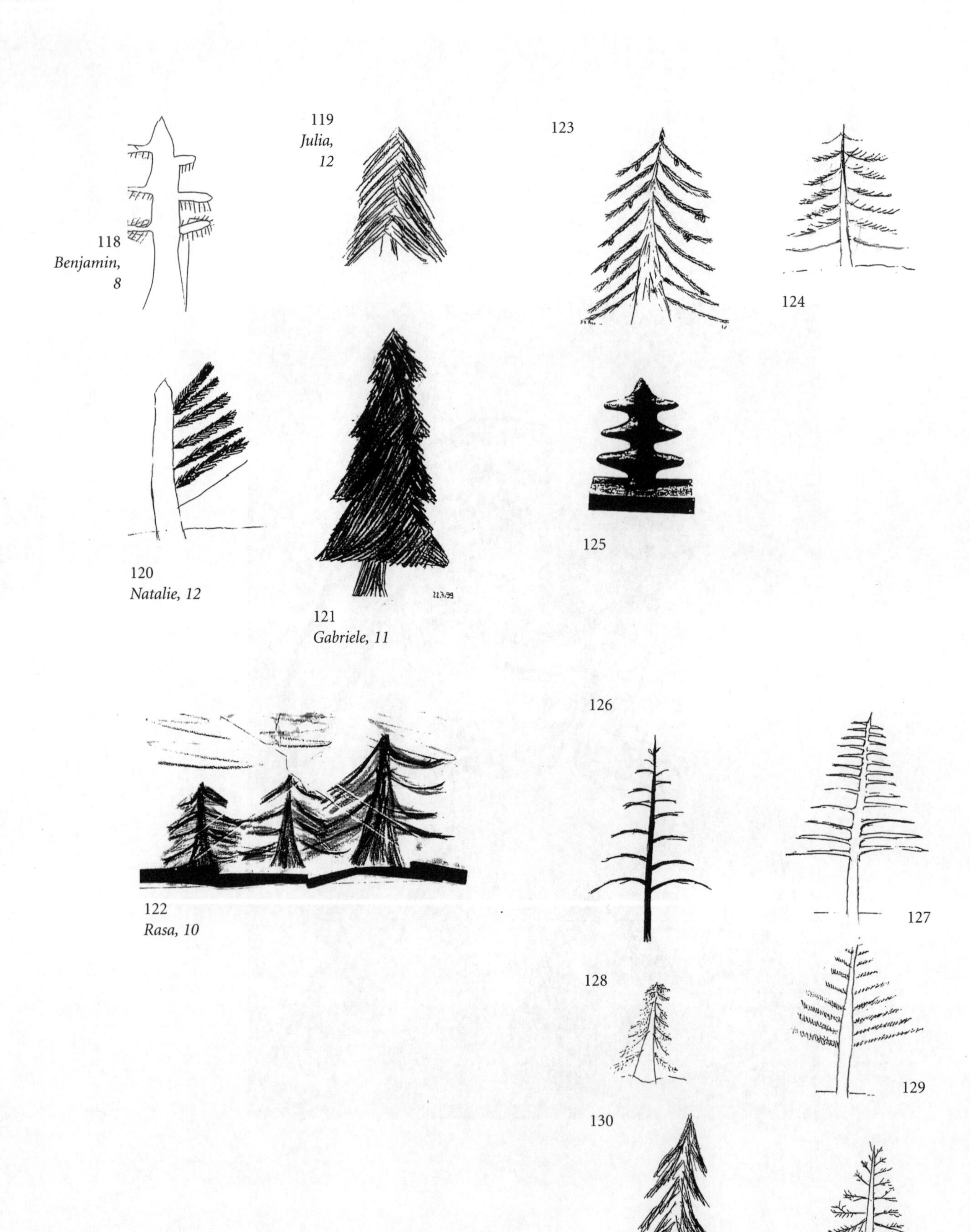

118 *Benjamin, 8*
119 *Julia, 12*
120 *Natalie, 12*
121 *Gabriele, 11*
122 *Rasa, 10*
123
124
125
126
127
128
129
130
131

132

133

134

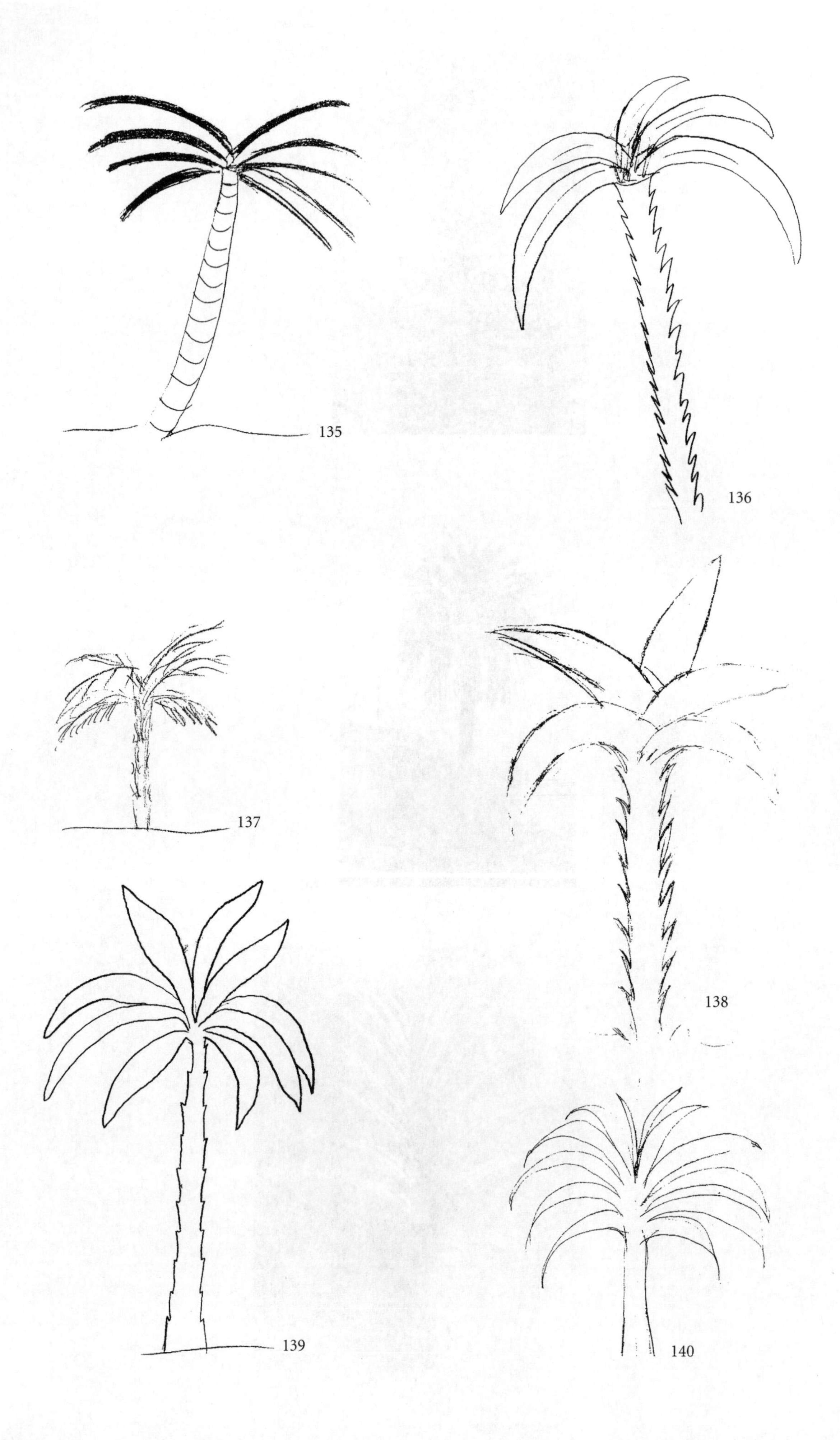
135
136
137
138
139
140

141

142

143

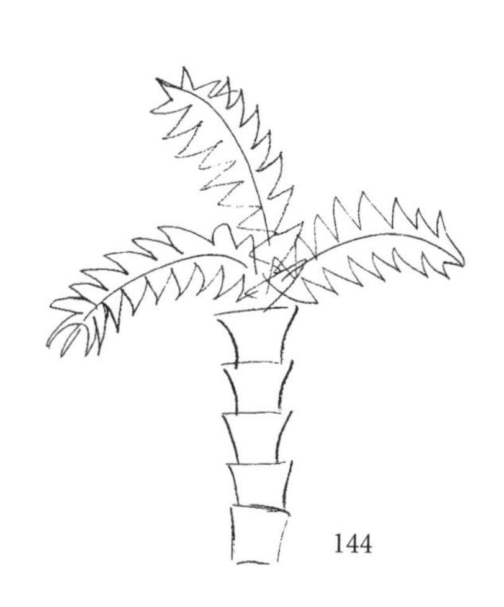
144

145

146

147
148
149
150
151
152
153
154
155
156
157
158
159
160

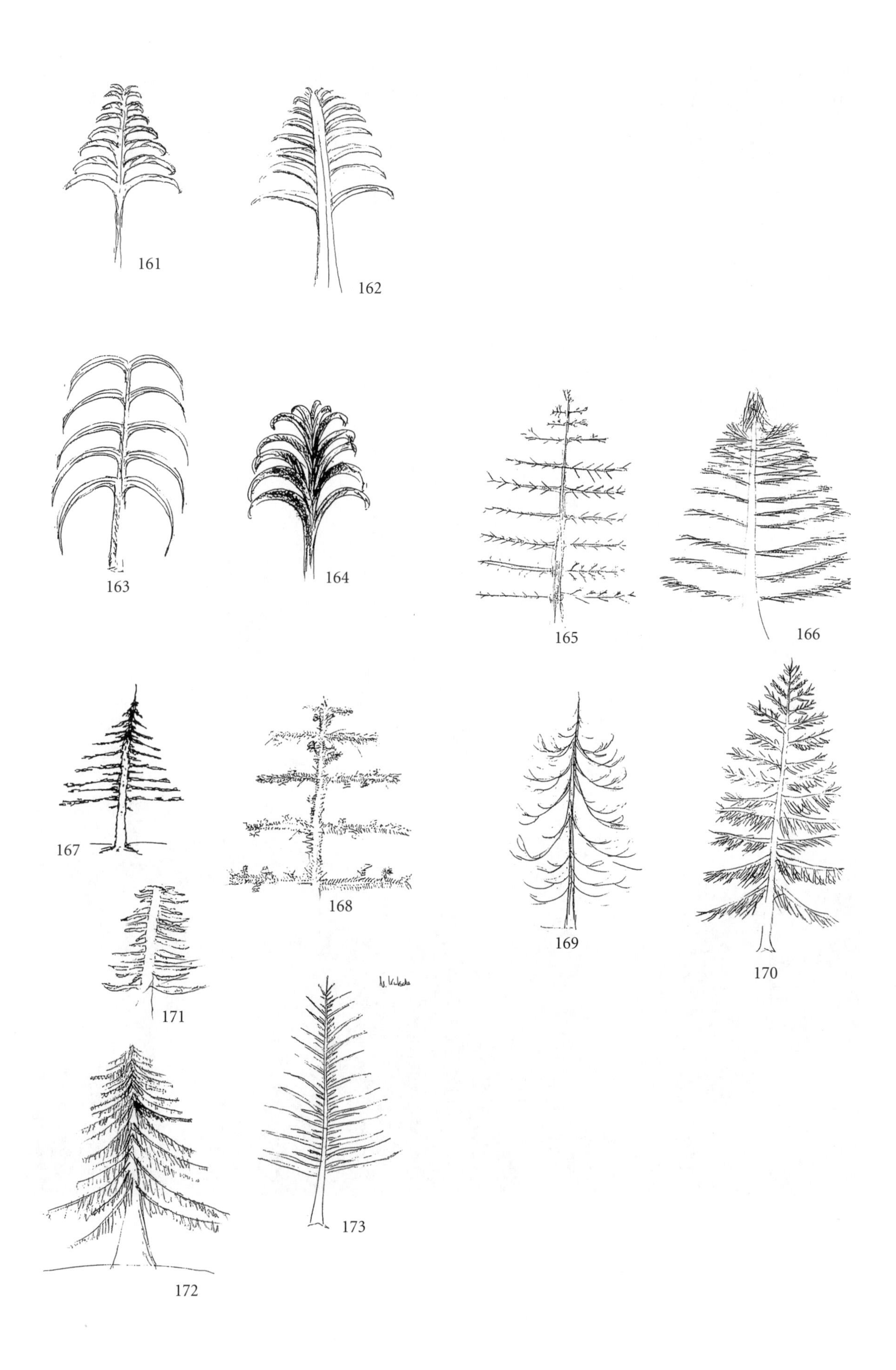
161
162
163
164
165
166
167
168
169
170
171
172
173

174

175

176

177

178

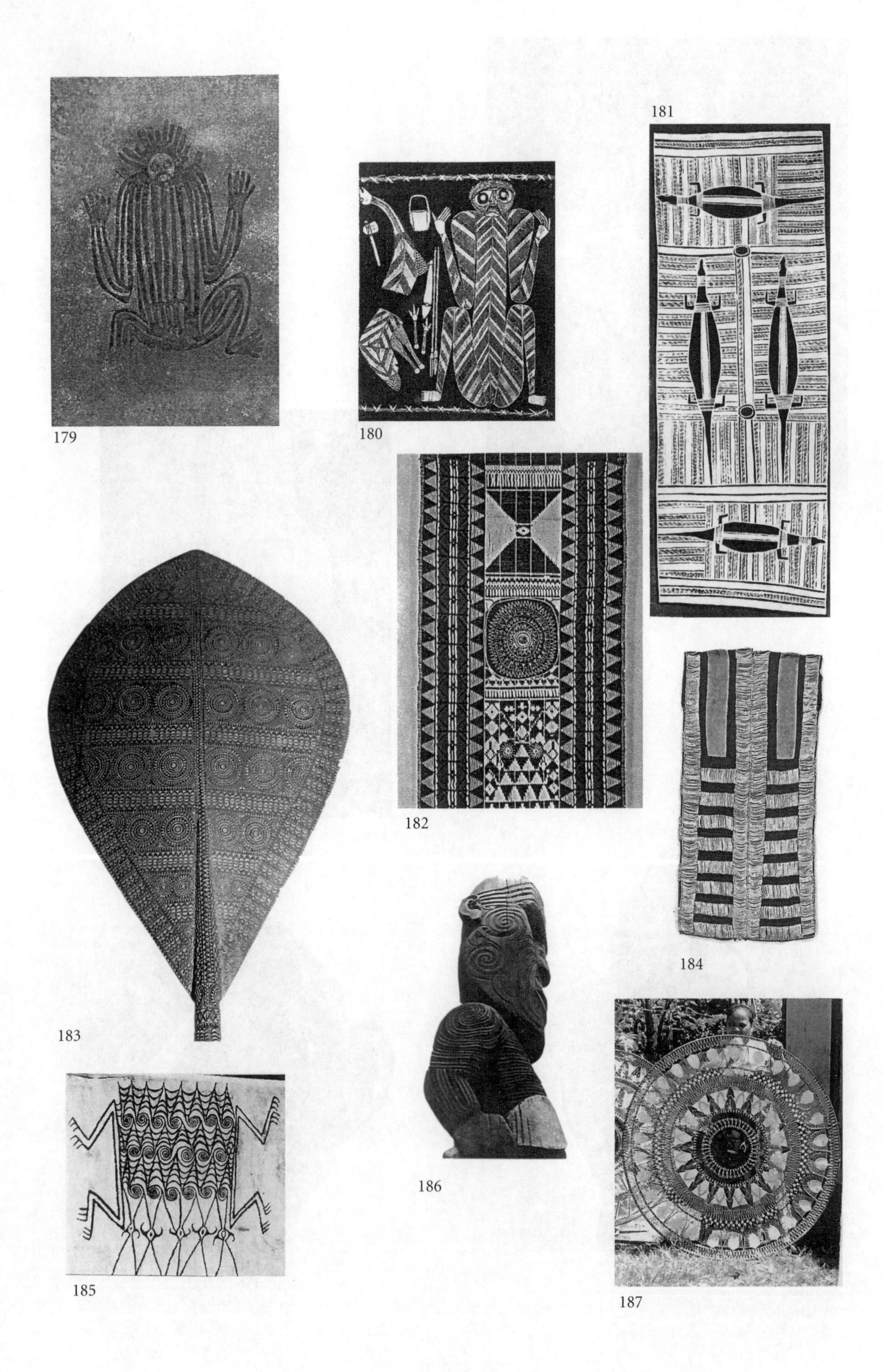
179
180
181
182
183
184
185
186
187

188

189

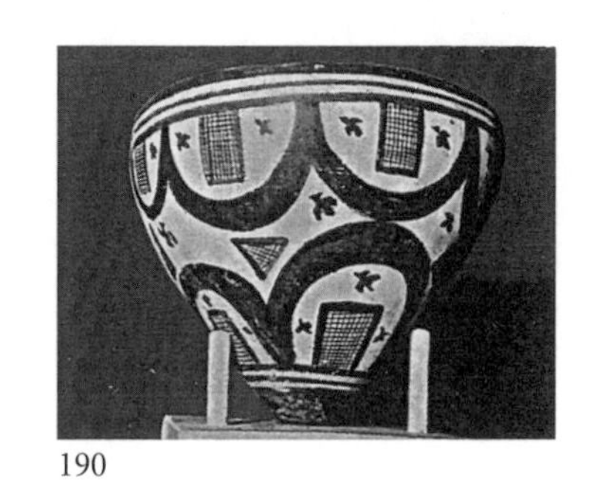
190

191

192

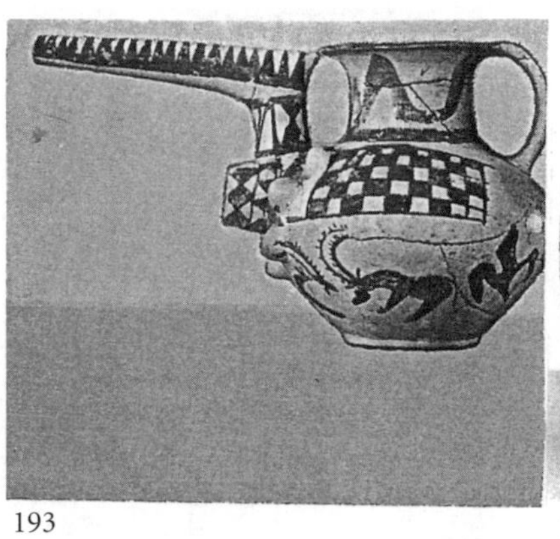
193

194

195

196

197

198

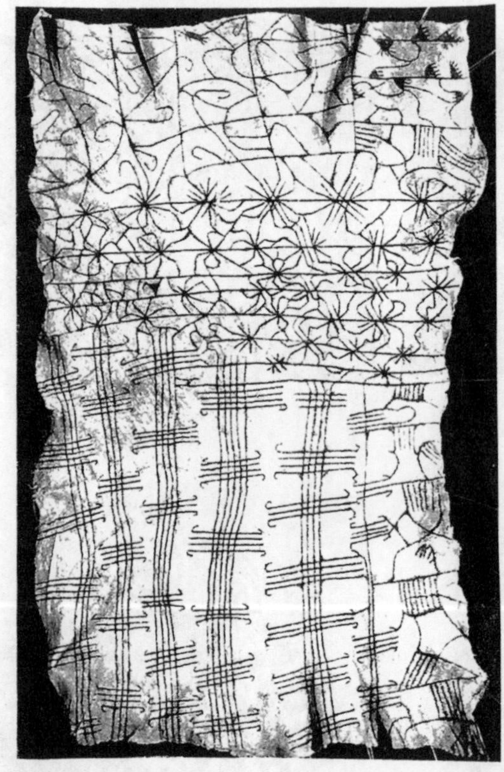

199

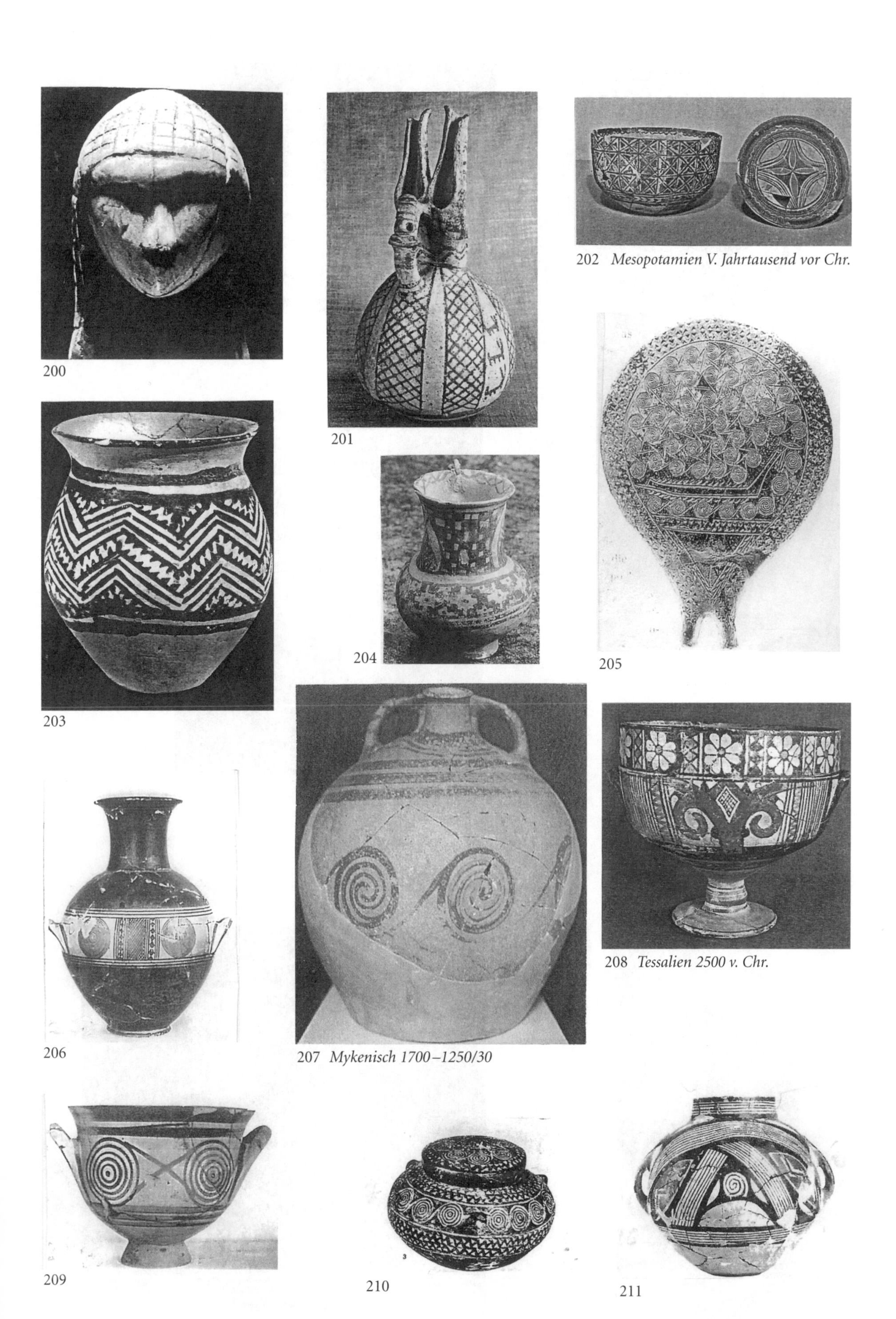
200

201

202 *Mesopotamien V. Jahrtausend vor Chr.*

203

204

205

206

207 *Mykenisch 1700–1250/30*

208 *Tessalien 2500 v. Chr.*

209

210

211

212

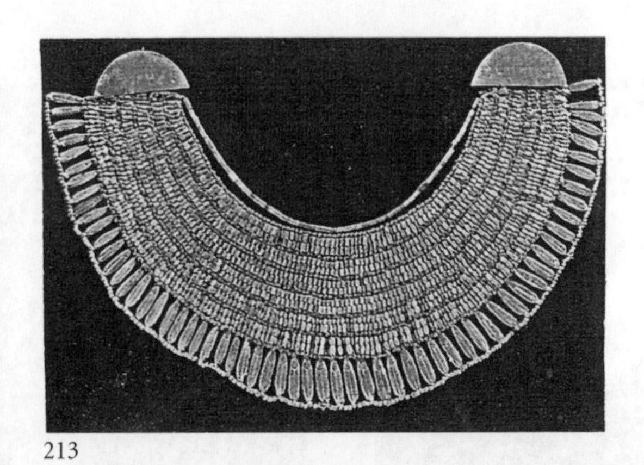
213

214

215

216

217

218

219

220

221

222

223

224

225

226

227

228

230

231

232

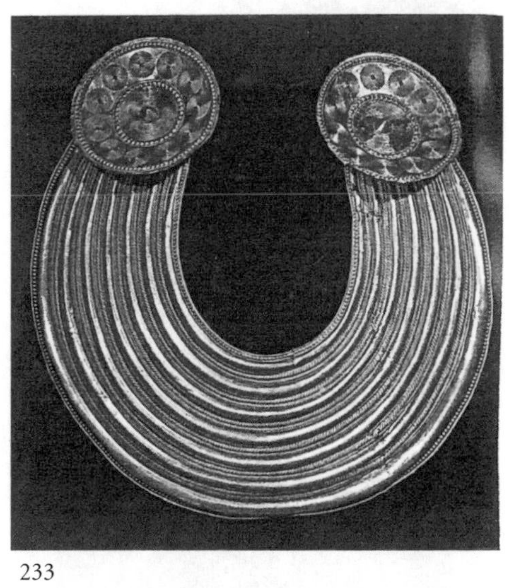
233

234

235

236

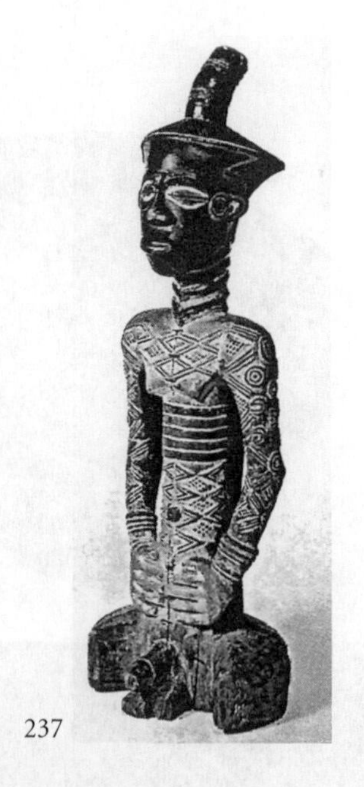
237

238

239

240

241

242

243

244

245

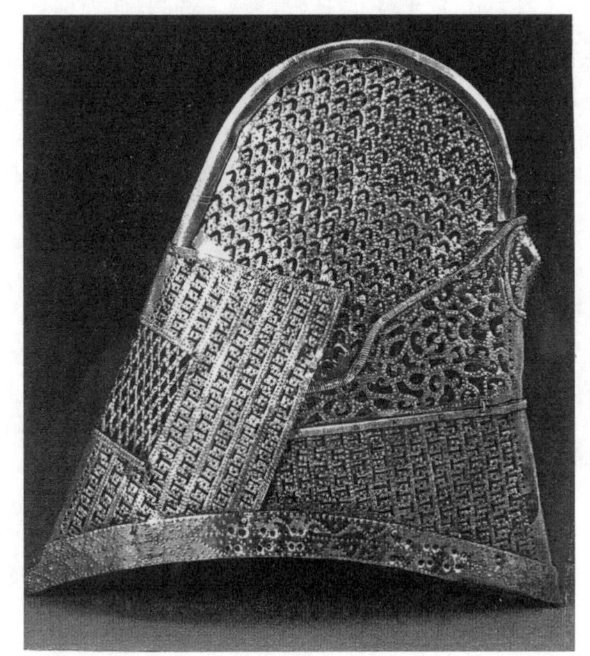

246

247

248

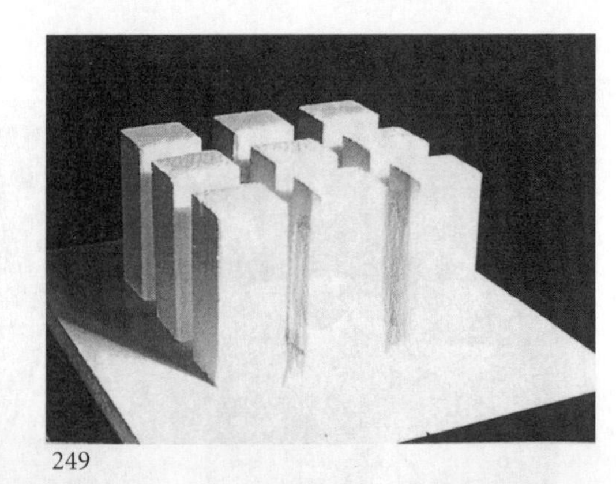

249

250

251

253

254

255

256

257

258

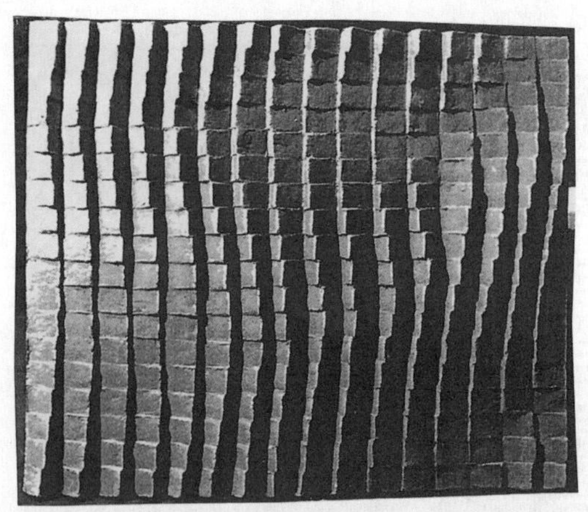
259

260

261

262

263

264

265

266

267

268

269

270

271

272

273

274

275

276

277

278

279

280

281

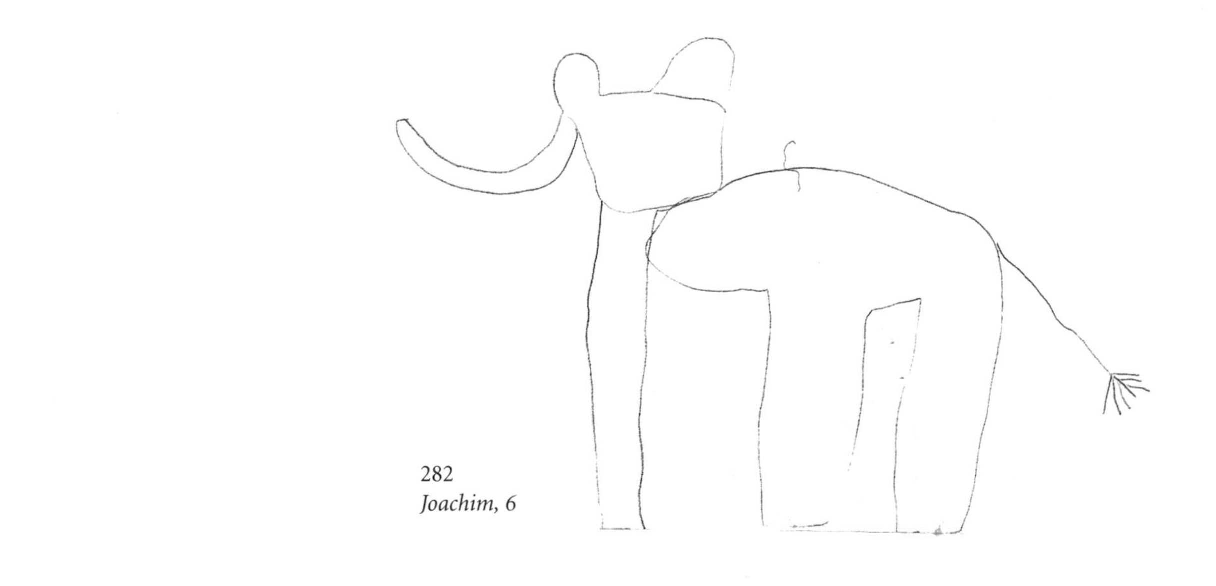

282
Joachim, 6

283
Nina, 8

284
Sabine, 11

285
Maria, 11

286
Rasa, 11

287
Natalie, 12

288
Olga, 13

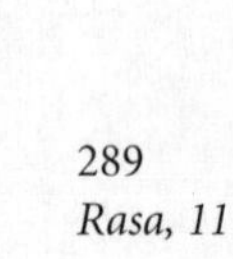

289
Rasa, 11

290
291
292
293
294
295
296
297

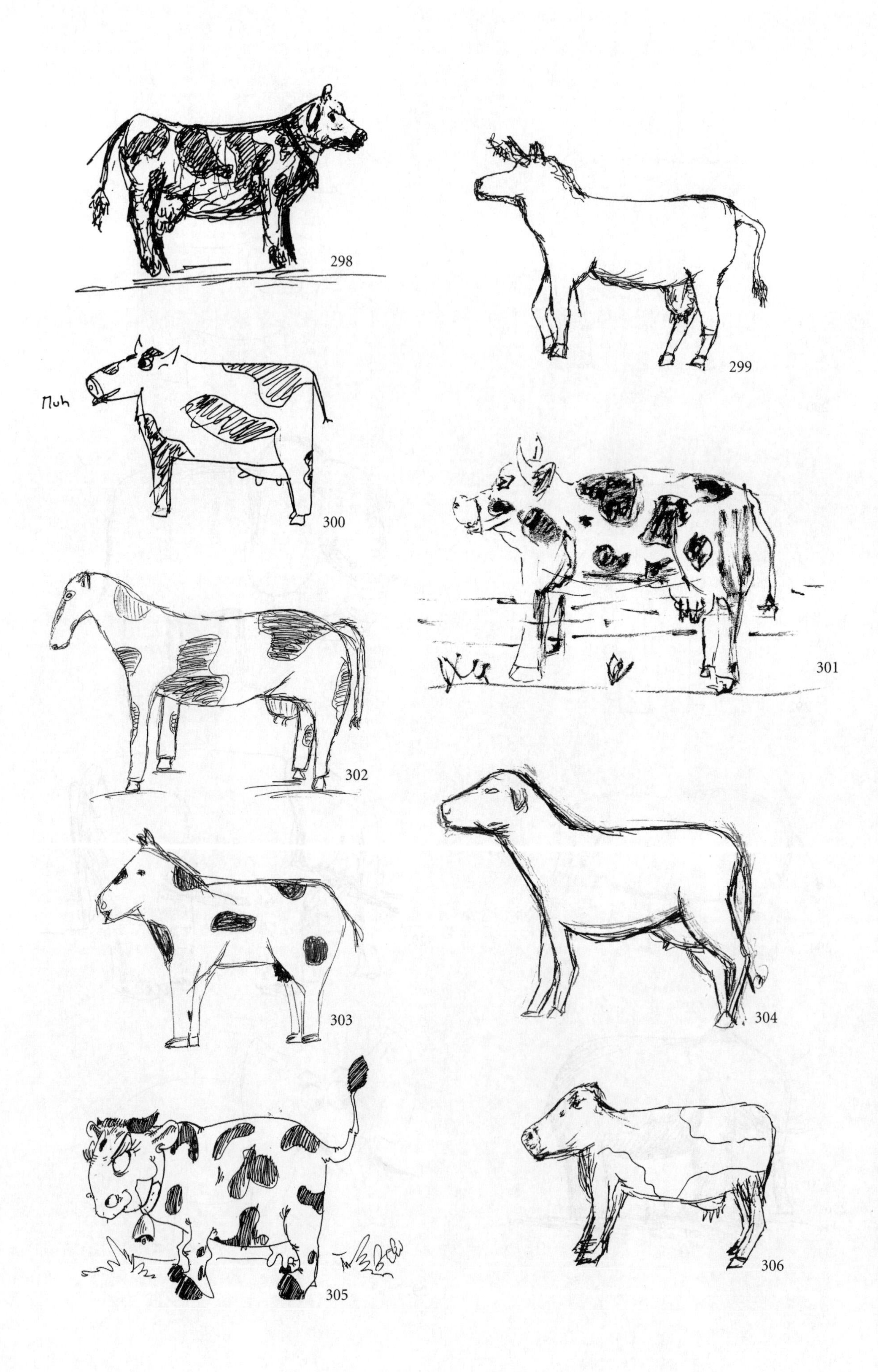
Muh
298
299
300
301
302
303
304
305
306

307 308 309 310 311 312 313 314 315

316
317
318
319
320
321
322
323
324

325 326 327 328 329 330

331 332 333

334 335 336 337 338 339 340 341 342 343 344

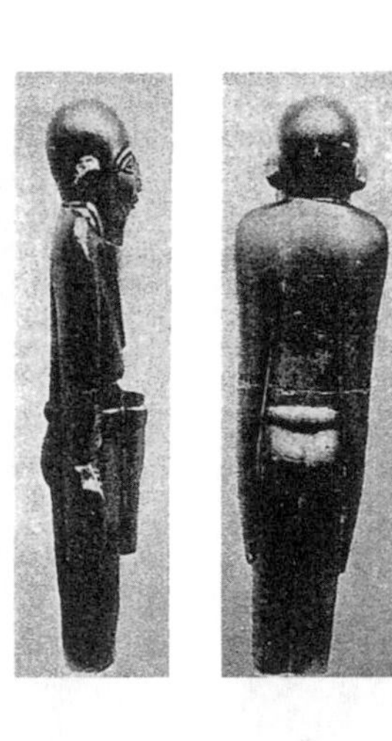
345

346

347

348

349

350

351

352

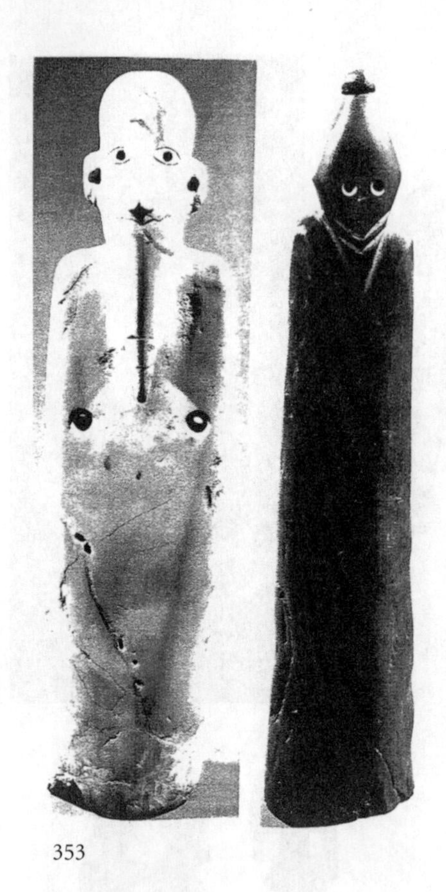

353

354

355

356

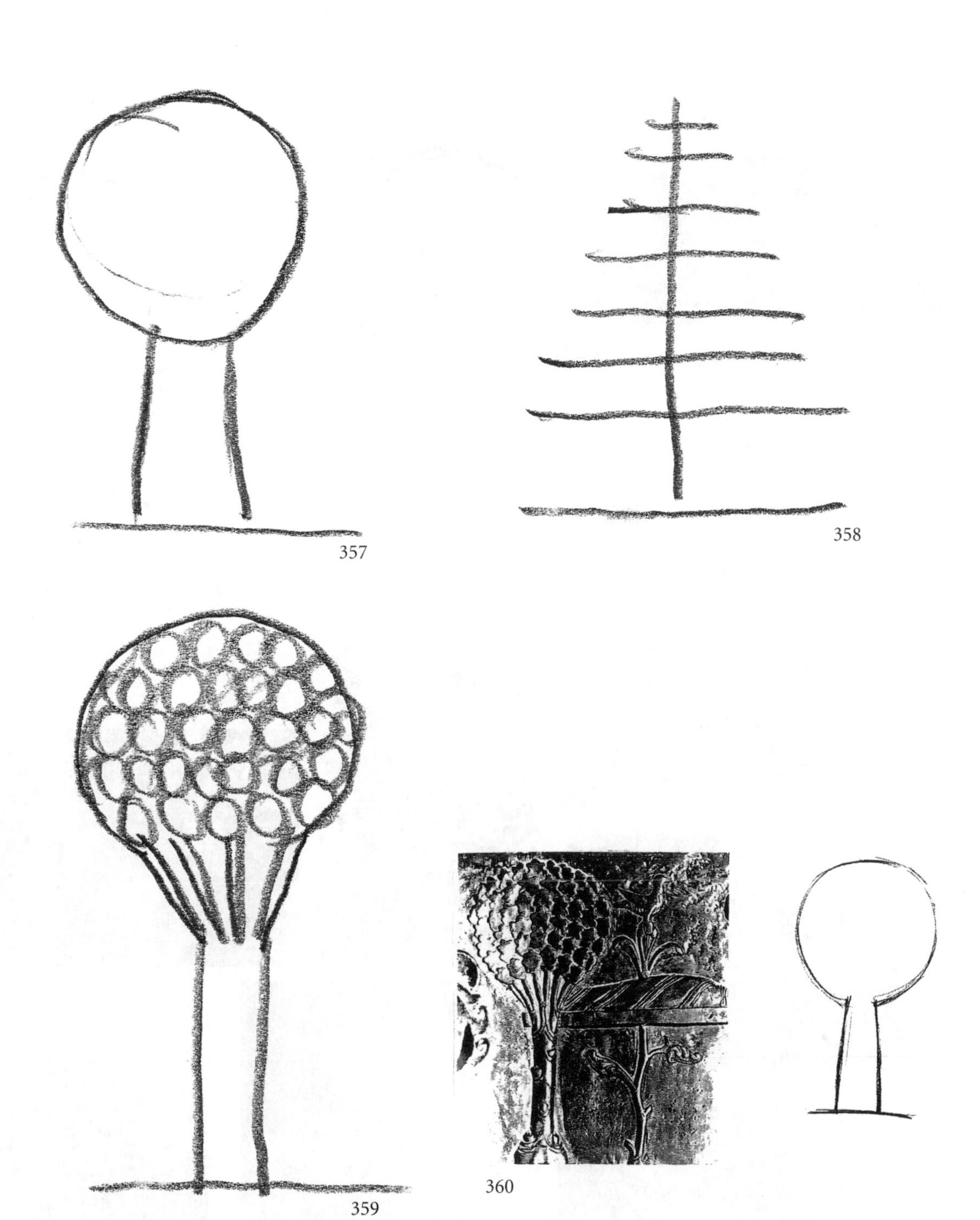
357
358
359
360

361
361 a
b
c
362
363
364
365
366
367
368
369
370

371
372
373
374
375
376
377
378

379

380

381

383

382

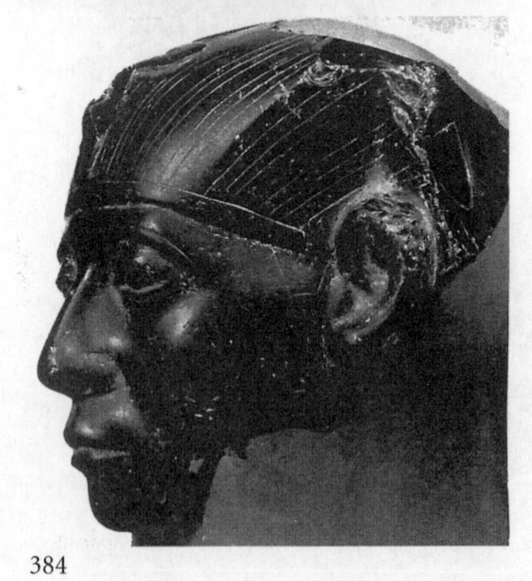

384

385

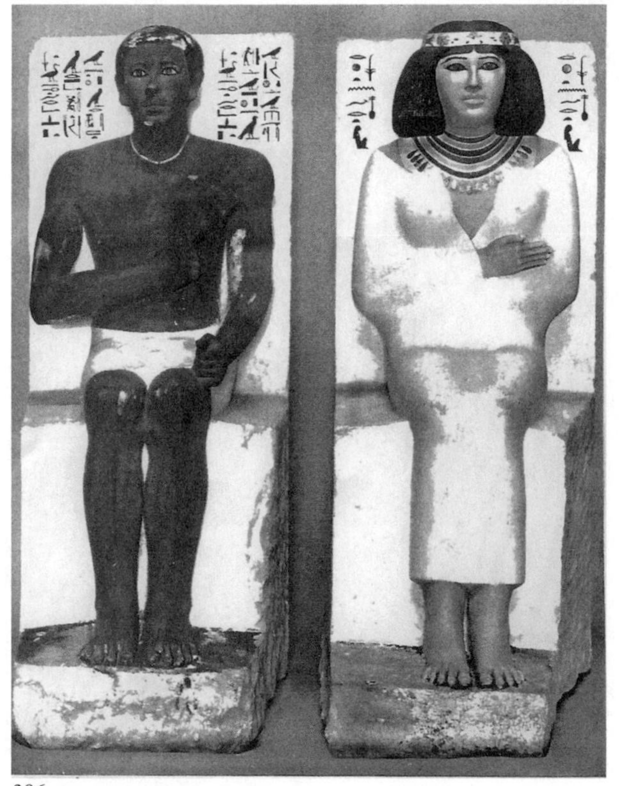

386

387

388

389

390

391

392

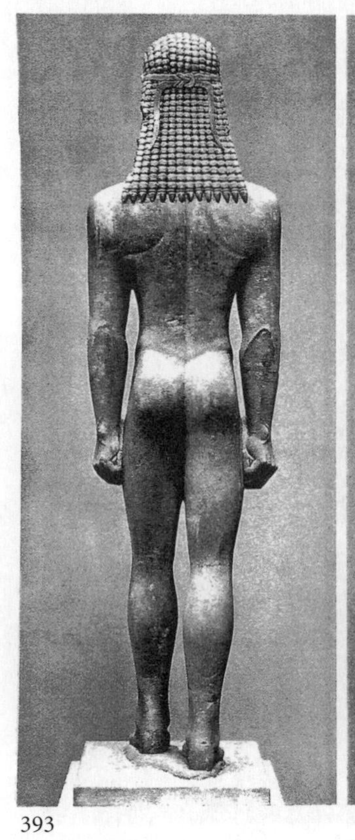
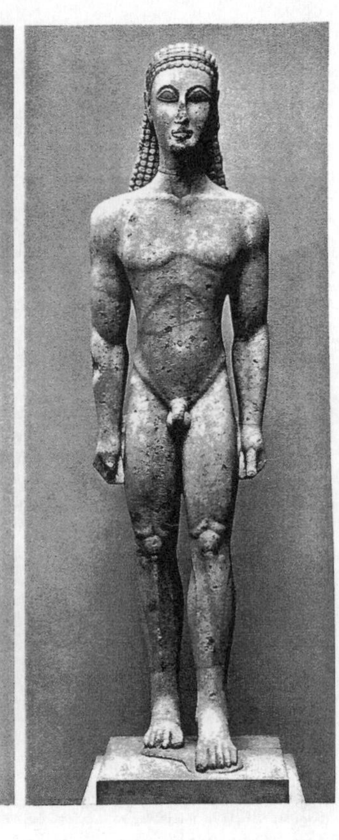

393

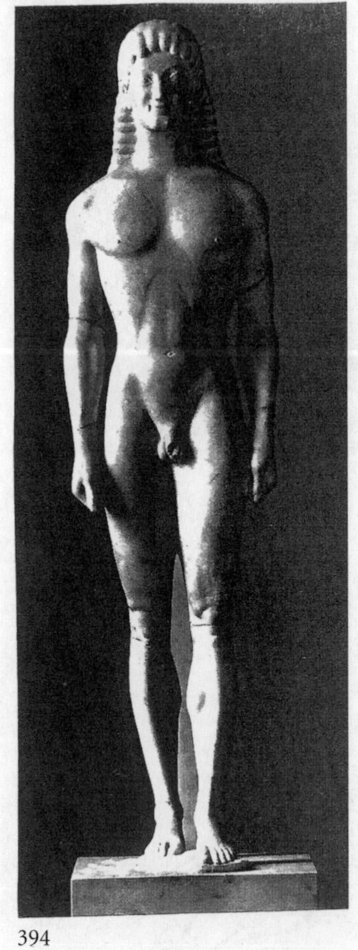

394

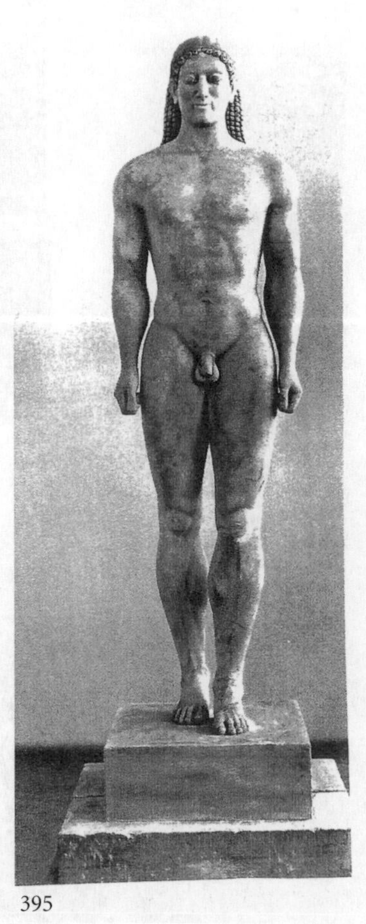

395

396

397

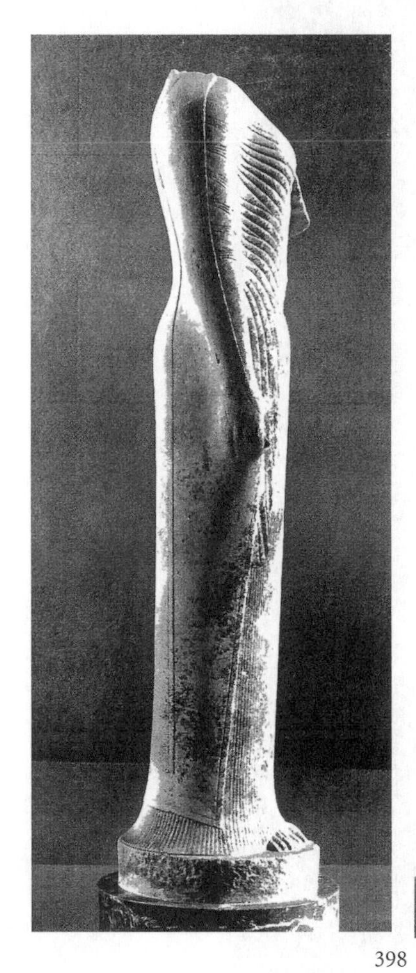

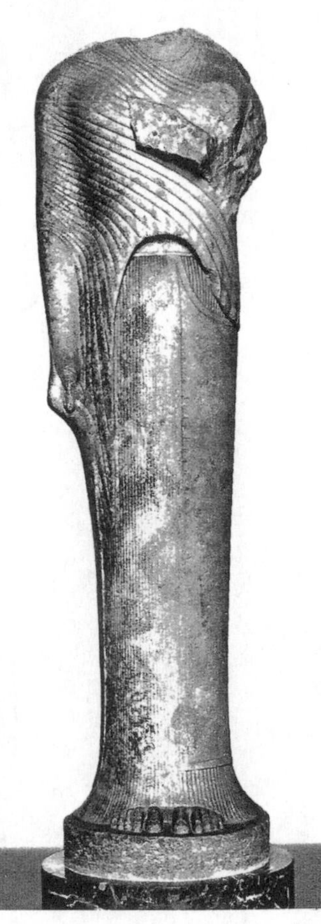

398

399

400

401

402 a

403

402

405

404

406

409

408

407

409 a

409 b

409 c

410

411

412

413

414

415

416
417
418
419
420
421
422

423
424
425
426
427
428
429
430
431

432
433
434
435
436
437
438
439

440
441
442
443
444
445

446

447

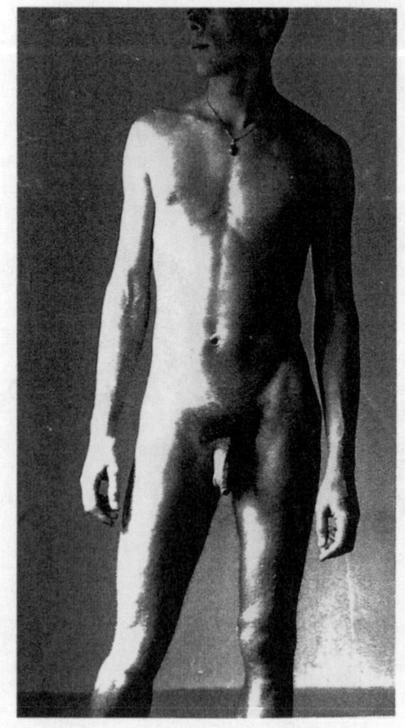
448

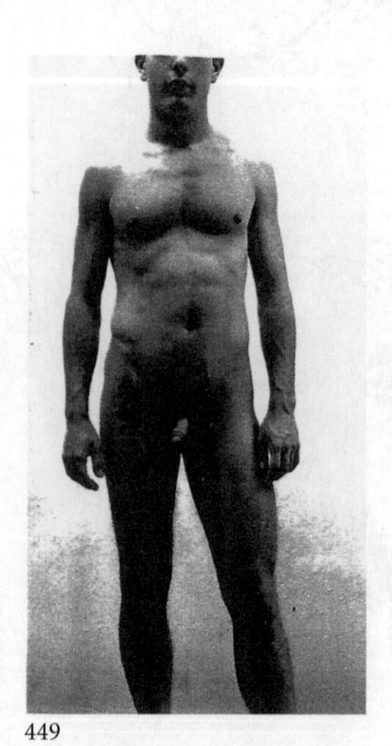
449

450

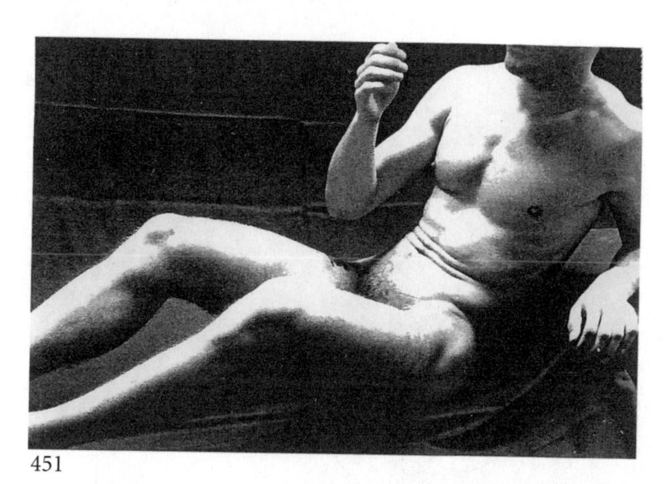

451

452

453

454

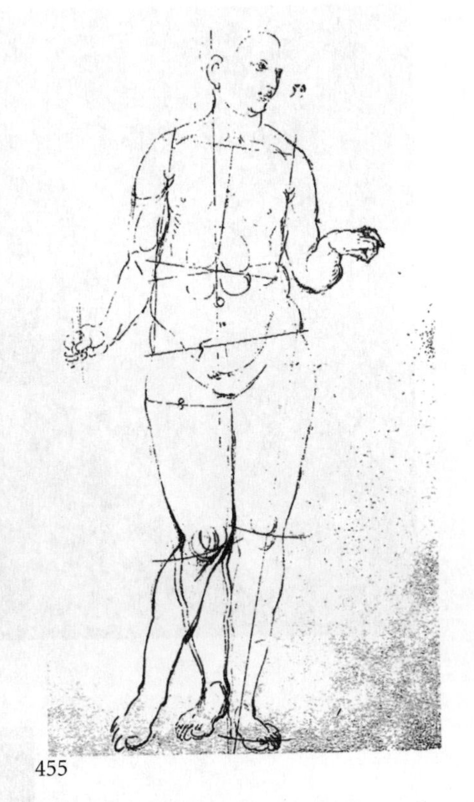

455

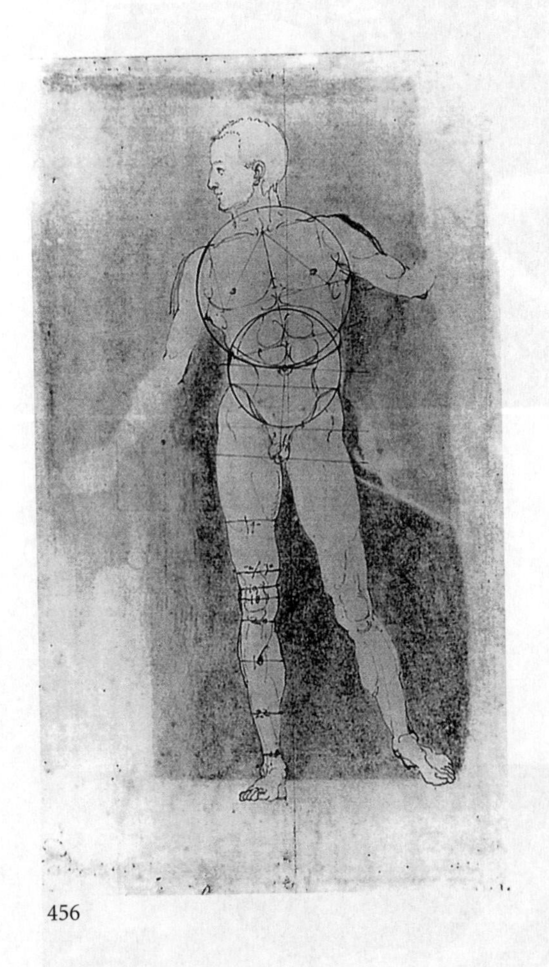

456

457

458

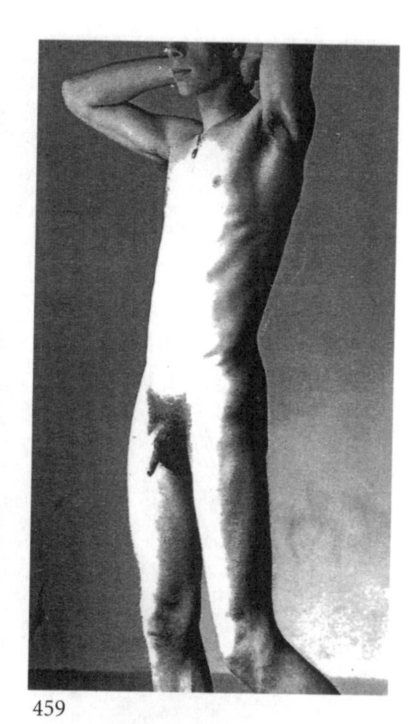
459

460

461

462

463

464

465

466

467

468

469

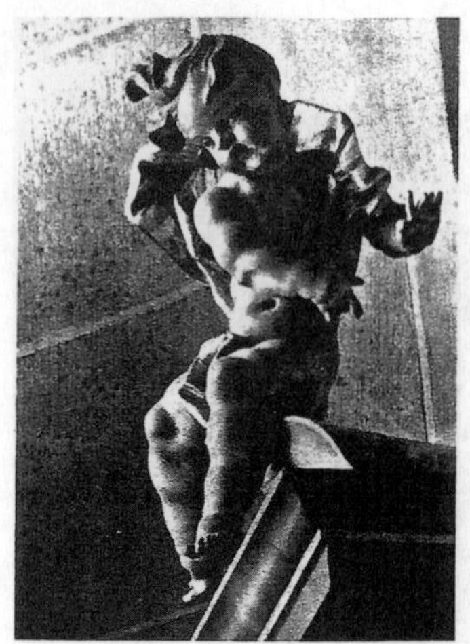

470

471

472

473

474

475

476 477 478 479 480 481 482 483

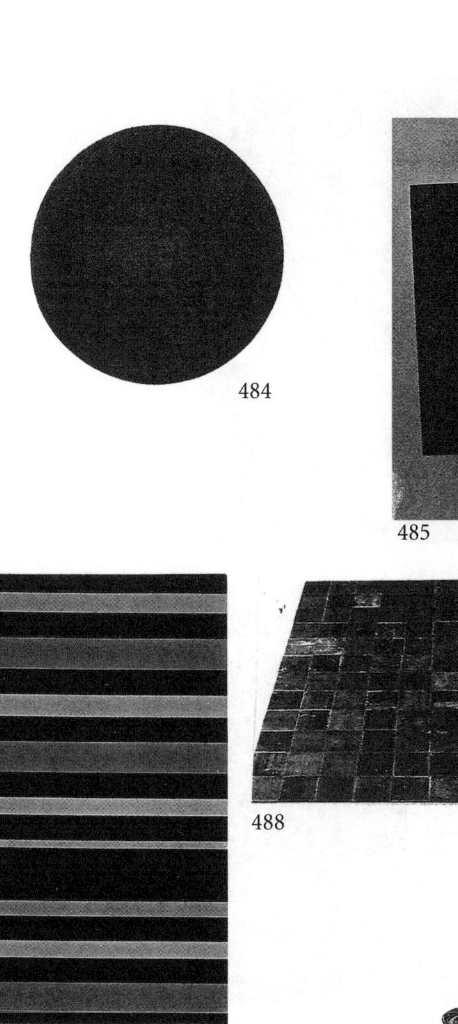

484

485

486

487

488

489

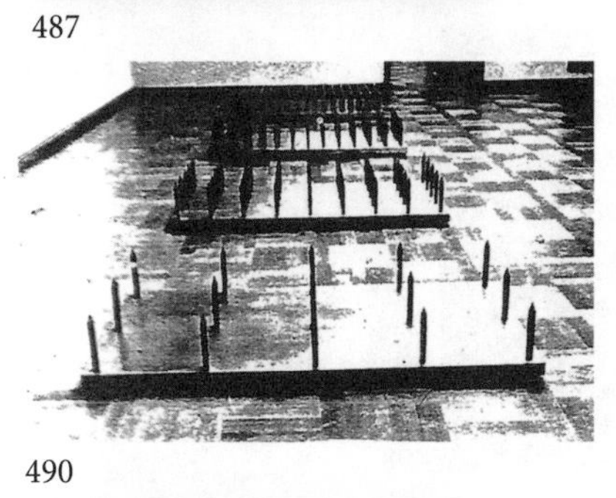

490

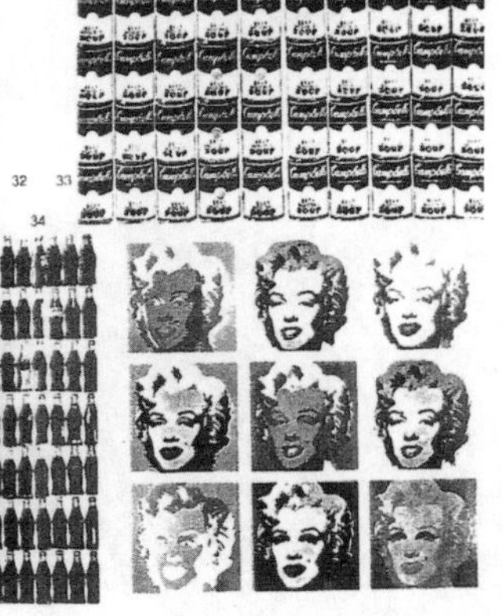

491

492

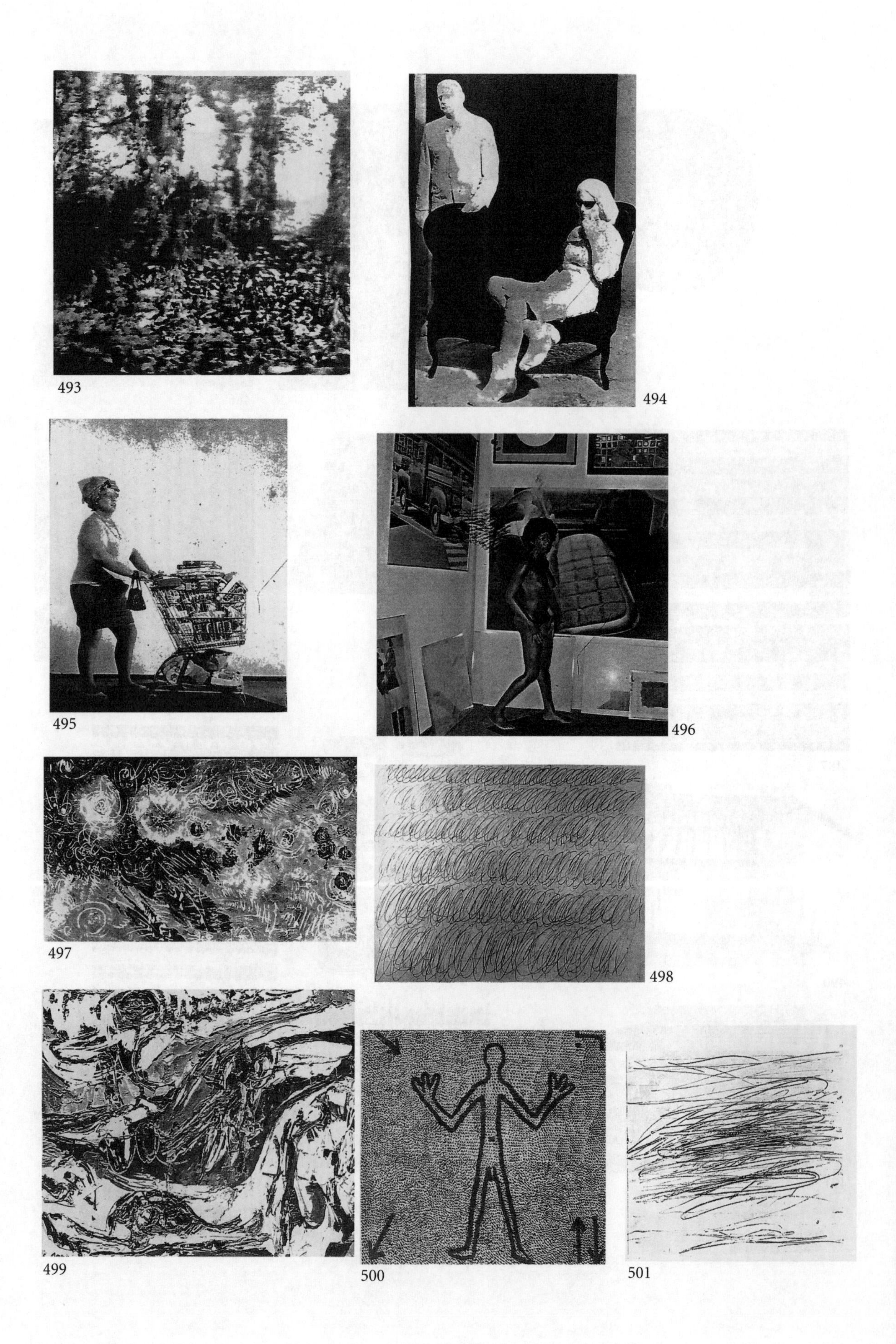

493 494 495 496 497 498 499 500 501

INDEX

Prof. em. Jürgen Weber
Institut für Elementares Formen
TU Braunschweig
Bevenroderstrasse 80
D-38108 Braunschweig

Printed in Germany

Printing: DruckVerlag Kettler, D-59199 Bönen

Printed on acid-free and chlorine-free bleached paper

SPIN: 10859053

With numerous (partly coloured) Figures

ISBN 3-211-83768-X Springer-Verlag Wien New York